Discard

Illustrated Treasury of

General Science Activities

Illustrated Treasury of

General Science Activities

Robert G. Hoehn

Parker Publishing Company, Inc.   West Nyack, N.Y.

© 1975 *by*

Parker Publishing Company, Inc.
West Nyack, New York

Library of Congress Cataloging in Publication Data

Hoehn, Robert G
   Illustrated treasury of general science activities.

   1. Science--Experiments.   I. Title.
Q182.3.H6   502'.8   75-14155
ISBN 0-13-451252-9

Printed in the United States of America

To Peggy, Valerie, Susan and James.

How This Book Will Help You Teach Science

This book provides science teachers with a large variety of stimulating demonstrations and experiments for classroom use. Each chapter includes activities and investigations for detailed study and observation. These activities help students carry out investigations on an individual basis with minimum guidance from the teacher. Equipment is inexpensive, readily available, and easily handled by the students. This practical guide works hand in hand with the teacher by:

* Providing a large reservoir of practical, interesting experiments, demonstrations, and activities which allow pupils to observe the science world.
* Arousing pupil interest and motivation. Many activities encourage friendly competition through contests, e.g., which student can prepare the tastiest seaweed dinner? Which student can build the strongest paper boat? (Chapter 6)
* Offering *Meal Wheel, Lion's Share,* and *Survival*— student games—which reinforce ecological principles. The teacher awards bonus points or small prizes to the winners. These teaching techniques increase student interest and involvement. (Chapter 10)
* Having students dissect the clam, squid, earthworm, crayfish, grasshopper, fish and frog in Chapter 9. Pupils explore the unique behavior of the living earthworm and crayfish.
* Emphasizing experiments which provide more than one correct answer.
* Encouraging students to make careful, accurate measurements and observations.
* Permitting students to work individually or in small groups. Pupils perform most activities in the classroom. However, many can be carried out at home or on field trips.

Every chapter includes meaningful, open-ended activities which can easily be supplemented with filmstrips, movies, slides, models, and overhead transparencies. You can encourage your students to pick an interesting topic, do the necessary research and make cassette recordings (including sound effects) which would offer additional value to the program. For example, a student-made tape that amplifies a variety of strange sounds will attract the attention of the most reluctant student. Homemade filmstrips, slides, and overlay transparencies have the same effect.

A teacher may choose from 10 science areas to select the material which best suits his teaching situation. For example, *Investigating the Sea and Its Inhabitants* (Chapter 6) suggests a student can actually test 6 factors which might influence the sinking of a ship. *Reaching Toward the Stars* (Chapter 2) offers recipes for making stony meteorites, metallic meteorites, tektites, mystery planets for student dissection, and plans for building a reflecting telescope from a shaving mirror.

Photographs, charts, and illustrations have been carefully selected to aid the teacher in preparing his daily lesson.

Each chapter lists additional problems, experiments, questions and investigations students may wish to solve. Because of the intriguing, on-going nature of many activities students will experiment further and expand many of these ideas into long-range projects.

ROBERT G. HOEHN

Acknowledgements

I express my thanks to Dr. Ruth Sherman for allowing me to share her *Chromobacter* unit.

A special thanks to Morton Handler, former Managing Editor, *Science Activities Magazine*, for his encouragement and delightful sense of humor.

To Clyde Bearden I owe thanks for his time in preparing the excellent photos which appear in the book.

Table of Contents

Unit Two
Exploring Life

UNIT ONE

THE RESTLESS EARTH

chapter one

17 ACTIVITIES THAT TEACH
THE ORIGINS OF LIFE

OVERVIEW

This chapter opens with a study of the earth's atmospheric gases. Students will investigate individual characteristics of hydrogen, oxygen, ammonia and methane gas.

We will cover protoplasm, the living material within an organism. The cell, basic unit of life, is reviewed. Life processes, including stimulus-response of several organisms, are also considered.

Students will examine the concept of spontaneous generation, simple forms of reproduction, and mitotic cell division.

In the closing section of the unit, UFO sightings and the possibility of life on other planets are discussed.

What is life?* We recognize organisms which possess similar physical and chemical structures as living things on earth. Some of these physical and chemical similarities include:

* Protoplasm—Living matter which makes up the living cell.
* Temperature limitations—Exceptions exist, but organisms usually live between 35° and 114°F.
* An atmosphere of nitrogen, oxygen, carbon dioxide, methane and water.

Begin this unit by mentioning how Drs. Alexander Oparin, Russian chemist, Harold Urey and Stanley Miller, American scientists, discovered that the amino acids, chemical forerunners of proteins, could have been formed in an atmosphere of methane, water vapor, ammonia, and hydrogen. The first organic substances, they

---

* Be sure students understand that scientists base their theories of the origin of life on laboratory findings. The activities offered in this section give students an opportunity to examine the elements thought to have existed in the primitive atmosphere. They are not intended to prove how life on earth began.

believe, were produced a few billion years ago in a primitive atmosphere made up of methane, water vapor, ammonia, and hydrogen.

Tell students the theory suggests that energy, perhaps lightning, caused reshuffling of chemicals which ultimately produced simple organic molecules. The final result—the plant and animal kingdom—may have sprung from these numerous, complicated chains of events.

ACTIVITY 1

How can a mixture of calcium, hydrogen and ammonium chloride produce ammonia?

Have students place 0.5 g of calcium hydroxide with the same amount of ammonium chloride in the palm of the hand (powdered form). Each of these separately is odorless. Then rub the hands together and bring the palms near the nose. Ask them to describe the odor.

Have pupils mix 5 g of ammonium chloride and 4 g of calcium hydroxide on a piece of paper. Place in a test tube, generator mouth downward (Figure 1). Use test tubes or gas collecting bottles to collect the ammonia gas by downward displacement of air. Cover with a glass plate or use a rubber stopper. Keep the room well ventilated.

Suggested Problems And Questions

1. Ask students to open a container of ammonia underwater. What does the water do? Does this indicate anything? Water rises quickly, demonstrating the great solubility of ammonia.
2. Tell students to insert a glowing splint into another bottle of gas held downward. Have them describe what happens. Ammonia gas will not support combustion.

ACTIVITY 2

How can hydrogen be produced?

Have students prepare hydrogen gas by placing zinc or iron pieces in a dilute hydrochloric or sulfuric acid solution. Let students experiment to see which combination produces the best results. Make sure students wear safety glasses. Have them collect the gas in tubes or displacement bottles.

Suggested Problems And Questions

1. Have pupils ignite a sample with a burning splint. What happens? A "pop" or barking sound occurs.

2. Ask them to bring a collecting bottle with a half air-half hydrogen mixture over a candle flame. A safe small explosion takes place. *Caution:* Remind students to wear safety glasses.

ACTIVITY 3

Have students collect the laboratory gas through water displacement, and test it with a burning splint. The gas will burn with a yellow flame. Natural gas is composed of approximately 90 per cent methane.
Caution: Pupils should wear safety glasses.

Demonstrations And Experiments

1. Fill a balloon or beach ball with a 2 to 1 hydrogen-oxygen mixture. Prepare a detergent solution in an aluminum pie pan or tray. Attach a rubber siphon-glass tubing apparatus (with pinch clamp) to the ball (Figure 2). Place the free end of the siphon below the surface of the detergent solution, release the pinch clamp, and allow the gas to form a large pile of bubbles. Scoop out a handful of bubbles and pass them over a candle flame.
    Caution: The loud explosion could be harmful to the ears. Tell students to hold their hands over their ears. For full effect, turn off the lights.
2. Show how water molecules break apart. Set up a standard Hoffman or H-type apparatus. Add 5 ml of concentrated sulfuric

Figure 2

Figure 1

acid to 150 ml's of water. Place this solution in the apparatus. Connect the wire electrodes to a 6-volt or a 12-volt battery or power source. Collect the gas by displacement. Hydrogen gathers in one tube and oxygen in another.

Insert a lighted splint into the mouth of the hydrogen tube. The gas gives off a characteristic "pop" sound. The sudden release of energy is followed by hydrogen combining suddenly with oxygen in the air to form water again. Show students the walls of the test tube.

3. Fill a small balloon with a hydrogen-oxygen mixture. Attach a three inch magnesium ribbon fuse to the end of the balloon. Light the fuse and cover your ears.

4. Find a 2-pound coffee can with a plastic lid. Punch a hole in each end of the container. Slowly fill the can with methane gas from the laboratory table. Insert the jet nozzle into one of the holes. When the gas escapes through the other end, turn the jet off. Tip or support the container at an angle. Be sure to keep the lid pointing away from the students. Strike a match and hold it over one of the holes. At first the flame burns bright, then slowly disappears. Seconds later a whooshing force blows the lid off.

Students may wish to do this experiment. If they do, have them wear safety glasses. Turn out the lights for full effect.

WAYS TO INVESTIGATE THE PHYSICAL PROPERTIES OF PROTOPLASM

Begin discussion by asking students to describe life. Guide them by asking what it is, where it exists, how it is important, and what determines whether something is alive or dead. Write their responses on the board. Bring up the term protoplasm. Tell them that protoplasm, all the living material that makes up an organism, consists of simple elements—carbon, hydrogen, oxygen, and nitrogen. These elements combine to form water, table salt, grape sugar, fats, oils, plus other substances.

ACTIVITY 4

Give students the opportunity to judge a "beauty contest." Prepare several "Miss America Mystery Capsules."* Fill each small container (capsule bottle) with from 12 to 20 ml of the following materials:

---

* What makes a Miss America? Well-distributed sugar, carbon, salt, protein, water—and fat.

Water—10 to 15 ml of tap water
Fat—1 ml of butter, margarine, or vegetable oil
Protein—1 ml of egg white
Salt—1 g of table salt
Grape Sugar—Sugar crystals, found near the
bottom of a jar of grape jelly
Carbon—Charcoal powder, briquet or coal
scrapings

Ask students to test this solution and analyze its contents. Allow them to taste, feel, smell, boil, evaporate, or test the solution in any way they wish. Stress safety factors in heating and boiling unknown solutions. Make sure all equipment is clean and free from chemical contaminants. Offer a prize to the student who correctly identifies the most items.

Suggested Problems And Questions

1. Have pupils design an experiment to investigate several ways to destroy protoplasm. Plants make excellent subjects. Suggest that heat, cold, acidic and basic soil, moisture, and light affect plant growth in various ways.
2. Water makes up from 50 to 96 per cent, by weight, of protoplasm. Have students compare the percentage moisture content of 2 leaves, young and old, from 2 different plants of the same specie and size. They can do this by:
    A. Weighing the leaf
    B. Extracting the moisture. Prepare a water bath by placing the leaf in a dry beaker, and then placing the beaker inside a larger beaker containing a small amount of water.
    C. Boil until the leaf becomes thoroughly dry
    D. Reweigh the leaf
    E. Divide the weight of Step A into the weight of Step D. Subtract this answer from 100. This will give the approximate amount of moisture loss.

Demonstrations And Experiments

1. Is protoplasm easy to describe? Starch—a white, odorless, tasteless, powdery carbohydrate—demonstrates the complexity of protoplasm. Mix corn starch and water together in a small bowl until you get a pasty substance. Ask 4 volunteers to do the following:
    Subject 1—Smell the substance, with eyes closed.

Subject 2—Quickly run your finger across the surface of the mixture, with eyes closed.

Subject 3—Slowly bob your finger up and down in the mixture, with eyes closed.

Subject 4—Remove a small portion of the mixture and roll into a ball, with eyes open.

Ask volunteers to describe the substance. Answers will vary.

2. Protoplasm is made up of inorganic (no carbon present) and organic (carbon present) materials. Ask students to collect items and group them according to organic and inorganic content.

Lipids, fatty substances, contain carbon and hydrogen. They are "greasy," an essential part of all protoplasm, and are stored or immediately used in the animal body. Have students test small pieces of bacon, ham, pork or beef for fat by soaking them in carbon tetrachloride for approximately 15 minutes. Keep the containers covered and the room well ventilated. Caution students not to inhale the fumes. Filter the sample and allow the solvent to evaporate under a heat lamp. Fat droplets remain in the dish. To calculate the percentage fat content, divide the weight of the original sample into the weight of the remaining fat. Multiply this answer by 100.

Suggested Problems And Questions

1. Have students test the percentage of fat in different foods—mashed potatoes, peanuts, walnuts, butter, and lard.
2. Ask pupils to report on the following questions:
   a. Does protoplasm carry the same chemical, physical, and biological properties in an amoeba, earthworm, and frog?
   b. How do terrestrial (land) animals keep from losing too much water? What is meant by metabolism?

PREPARING PLANT AND ANIMAL CELLS FOR MICROSCOPIC STUDY

Most students have previously studied the living cell. However, the following discussion and activity will serve as a brief review of important concepts.

Tell students that cells, the building blocks of life, serve as a protoplasmic unit necessary in carrying on life's activities. Show them diagrams or models of both plant and animal cells. Point out the general difference between the two, emphasizing the animal cell has a round shape in contrast to the "box like" shape of the plant cell. Also, point out that the animal cell has a cell membrane which is less distinguishable than the cell wall of the plant cell. Stress that living cells are made up of protoplasm and all, with the exception of certain blood cells, contain a nucleus and cytoplasm.

ACTIVITY 5

Have students make slides of thin sections of cork, onionskin cells, and cheek cells scraped from the inside of the mouth with a toothpick. Tell pupils to prepare slides by:

* Adding a drop of water to the center of a slide
* Placing the cell material on the water drop
* Covering with a cover slip
* Putting a drop of dilute iodine at the edge of the cover slip. This stain will make cellular structures easier to see.

Ask students to make sketches of the different cells under high and low power.

Suggested Problems And Questions

1. Have students find the function for the following structures: cell membrane, cell wall, cytoplasm, and nucleus. Ask students to report on how these structures interact with each other to produce an effective unit.
2. Divide the students into groups and have them report on these questions:
   a. What do amoebas and cheek cells have in common? How do they differ?
   b. Which parts of the cell are living? Which parts are considered non-living?
   c. Why is the nucleus believed to control the cell's activities?
   d. What is the purpose of the cytoplasm?
3. Ask pupils to make plant or animal cell models. Offer these hints:
   a. Cut a large Styrofoam or plasticine clay ball in half. Insert pins with colored plastic heads throughout the model to illustrate cellular structure.
   b. Fill a Petri dish with clay. Use different colored chemicals— sulfur, copper sulfate, salt, carbon powder, potassium dichromate—to show cell parts.
   c. Fill a Petri dish with a pasty plaster of Paris mixture. Mold the desired structures. When the plaster dries, paint the cellular structures.
   d. Find out who your creative students are. Give them clear plastic containers, clay, Styrofoam pieces, pins, pipe cleaners, boiling chips, filter floss, cotton, thread, glass or plastic beads, rice,

beans, popcorn kernels, and paper. Have them devise their own cell models.

Challenge the class by asking them to name the active processes associated with living organisms. Center the discussion around the following concepts:

* Living organisms use food
* Living organisms use oxygen
* Living organisms excrete waste products
* Living organisms react to internal and external stimuli
* Living organisms reproduce
* Some living organisms grow and move about

Bring up the topic of viruses. Ask students if viruses are alive or dead. Tell them scientists are in disagreement concerning this point. Viruses appear to use food and oxygen to produce energy. They seem to grow and multiply only inside certain cells. Emphasize that viruses, growing in living cells, produce diseases such as tobacco and tomato leaf mosaic, rabies, small pox, polio, mumps, influenza, and yellow fever.

ACTIVITY 6

Give students some plasticine clay. Show them diagrams and illustrations of different shaped viruses—spherical, oblong, spindle, cubical, and spiral. Have them make representative models. These finished products, placed in plastic containers, make impressive displays.

Suggested Problems And Questions

1. Ask students to collect plants infected by virus. Tobacco leaves, some potatoes, lettuce, and cucumbers are possibilities. Have them investigate the damage and determine what percentage of the plant seems destroyed. (If these plants are unavailable, have students check farm or agricultural journals for information and photographs.)
2. Have pupils find out which animals carry virus and transmit disease.
3. Tell students to answer the following questions:
   a. Are all viruses harmful? Not all viruses cause disease. Some are passive and do not bring about death or serious disorder to the host organism.

b. What is the chemical make-up of viruses? All viruses consist of a protein unit molecule called RNA (Ribonucleic Acid).

c. What is an electron microscope? How is it used to investigate viruses? An electron microscope is an extremely powerful microscope capable of making viruses visible. The wavelength of electrons emitted from the electron microscope are small enough to show viruses.

Demonstrations And Experiments

1. Do the following mock life demonstration on an overlay transparency projector. Don't let students see you prepare the chemicals.

   Pour a one quarter inch layer of dilute nitric acid into the bottom of a glass tray. Add 2 or 3 drops of mercury into the acid. Place some potassium dichromate crystals next to the mercury. Note: You may have to coax the mercury with a probe. Also, you may need to add additional crystals to keep the mercury active. Ask students the following questions:

   a. Does this experiment demonstrate life processes? If so, which ones? Students will probably say the mercury is alive because it moves, eats, and grows.

   b. Does the "organism" appear to move on its own? Students will generally agree that it does.

   c. Do the larger particles appear to feed upon the smaller ones? Occasionally a larger piece of mercury does seem to devour a smaller one.

   d. What conclusions can you make about this demonstration? Answers will vary. When you complete the demonstration, be sure to show this "organism" is not alive.

ACTIVITIES 7-8

Observing a Three-Day Yeast Experiment.

Have students study yeast cells for the next few days. Tell them yeast is a one-celled plant which cannot make its own food and contains no chlorophyll (the green coloring material found in green plants).

Give each group of students these directions:

* Label 5 gas collecting bottles or flasks (A through E).
* Mix 50 g of sugar with 450 ml of warm water (100°F).
* Put 100 ml. of the solution in each of the containers.

Prepare each bottle as follows:
1. Bottle A—Sugar solution only
2. Bottle B—Sugar solution only
3. Bottle C—Sugar solution with grass, dirt, and soap added.
4. Bottle D—Sugar solution and 20g of table salt
5. Bottle E—Sugar solution and food coloring.

*   Add 0.5 g of powdered yeast to containers A, C, D, and E. Add nothing to container B. Tightly cover the mouth of each container. Set them aside.

For the next 3 days have students prepare microscopic slides from each bottle. Ask them to observe each slide under high and low power of their microscope. *Caution:* Use a different dropper for each container. Remind pupils to occasionally shake their containers.

Have students answer the following questions:

First Day Observation

1. Sketch the yeast cells. What shape are they?
   Yeast cells are generally oblong, egg shaped, or rounded.
2. Do they appear to move? If so, how?
   No, yeast cells are stationary.
3. What color are yeast cells?
   No particular color. Mostly transparent.
4. Count the yeast cells. What is an easy way to do this? Count those found in a quarter section of the slide. Multiply that number by 4.
5. How do you think yeast cells reproduce? Can you see evidence of this?
   They reproduce by budding. Some may be in the process of reproduction.
6. Are any other organisms present? If so, try to identify them. Make a sketch of these organisms. Answers will vary.
7. Do yeast cells clump together or remain separated?
   They generally stay together.
8. How many life processes can you observe? Name them.
   Answers will vary.
9. Do any of the yeast cells appear to be feeding?
   Answers will vary.

Second Day

Repeat microscopic observations. Answer questions 4, 5, 6, 7, 8, and 9.

Third Day

Repeat microscopic observations. Answer questions 4, 5, 6, 7, 8, and 9. Do any of the yeast cells appear to be dying? Are the containers becoming cloudy? If so, which ones? How do you account for this? Answers will vary.

Suggested Problems And Questions

1. Ask pupils to graph the daily yeast cell count. By what per cent did the yeast cell population increase from Day 1 to Day 3?
2. Did Bottle B contain yeast cells? If yeast cells were present, how would you account for this?
3. Tell students to set the bottles aside for one week. After that time check the bottles. What changes have taken place? Did the yeast cell population increase, decrease, or stay the same? Explain your answer.
4. Have students report on how the baking and alcohol industry use yeast.

Demonstrations And Experiments

1. Some students may wish to test yeast cell reaction to different temperatures, bright light, darkness, and weak acid solutions.
2. Yeast plants bring about a chemical breakdown of sugar, releasing the energy which they need for their life processes. This type of chemical breakdown, which is also caused by many other microorganisms, is called fermentation. One product of fermentation by yeast is a gas.

    Show students how yeast cells give off carbon dioxide. Mix yeast with a sugar solution (10%) and place in a gas collecting bottle. Connect the bottle to a beaker or test tube of clear limewater. Point out the bubbles rising in the collecting tube contain carbon dioxide gas. Carbon dioxide turns limewater cloudy.
3. Do yeast cells prefer one food substance over another? Have students test this problem by placing 0.5g of yeast in 7 different bottles containing a 10 per cent solution of corn syrup, table sugar, honey, brown sugar, powdered sugar, and molasses.
4. Can yeast cells survive in a saccharin solution? Ask students to determine how they might find out. (Saccharin contains sodium saccharin but no protein, fat, carbohydrates, or calories.)

HOW DO EARTHWORMS RESPOND TO DIFFERENT STIMULI?

Begin this section with a classroom demonstration. Ask your students to think about a dentist's drill as it burrows deeper and deeper into a sore, sensitive tooth. Students will have little trouble identifying the stimulus from the response.

Mention that living organisms, in varying degrees, respond to stimulation. The organization and capacity of the nervous system allows a living substance to be aware of its environment and respond accordingly. Their response, in many cases, aids in their survival.

ACTIVITY 9

How do earthworms respond to light? Give students enough clay and glass slides to build a maze (Figure 3). Blacken both sides of two slides over a candle flame. Place the maze in a tray covered with moist paper. Place a worm, head end first, into the maze entrance. Set a light source directly over the maze. Caution: Do not place light too close to glass slides.

Do this experiment with several different worms which live in a dark, moist area. Place the worm in the maze, turn off the room lights, and turn on the overhead lamp.

Suggested Problems And Questions

1. Have pupils test other organisms—sowbugs, millipedes, centipedes, ground beetles, crickets—for light sensitivity.
2. Ask students the reason for using several different subjects instead of only one. If only one animal is used, habituation may occur.
3. Ask students to test what happens if you place an earthworm tail first into a maze.
4. Do earthworms react the same way to different colored lights? Give students red, blue, and green light bulbs of the same intensity. Have them test worm reaction to each light.

ACTIVITY 10

Students will enjoy playing "pick-a-niche." First have them divide a Petri dish lid into 4 equal parts giving each part a different number (Figure 4). Sections 1 and 3 should be covered with black paper or darkened with a grease pencil. Then have them build compartment walls from clay, leaving the center area open (See Figure 4).

Mix up the following ingredients for each compartment:

> Compartment 1—Moist soil, bits of leaves and
> grass, and small pieces of earthworms.
> Compartment 2—Coffee grounds and sand.

Figure 3

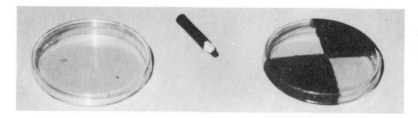

Figure 4

Compartment 3—Moist soil containing a 10 per cent salt solution.
Compartment 4—Dry soil

Ask students to fill their dish compartments with the appropriate mixture. Do not tell students what ingredients went into each mixture. Have them place a small centipede in the center of their dishes. Ask them to make two guesses: Which compartment the centipede will choose* and the ingredients which made up the selected compartment.

Suggested Problems And Questions

1. Ask students to test whether or not light intensity, temperature, or texture effects which compartment a centipede selects.
2. Have pupils test the environmental factors which might influence what compartment a sowbug, millipede, or ground beetle would prefer.

---

* Centipedes, if not fed regularly, will turn cannibalistic and eat one another. Hungry centipedes favor small insects and pieces of earthworms.

3. Ask pupils to suggest reasons why there might be variation of behavior in insects of the same specie.
4. Find out which centipedes locate a niche faster—hungry or well-fed ones.

ACTIVITY 11

Give students several earthworms, beakers, and tuning forks or other instruments for producing substratum vibration. Tell them to place a moist paper pad on the bottom of the beaker, insert an earthworm, and strike the fork with a rubber bung (a piece of glass tubing inserted into a rubber stopper makes a satisfactory bung), and touch the top of the beaker. Continue touching the beaker with the fork at intervals. Have them list every behavior they observe (withdrawal, twisting, movement toward or away from the vibration, etc.).

Suggested Problems And Questions

1. Have pupils test the reaction of various insects to vibration. Encourage them to use different kinds of devices such as finger thumping or an electrical buzzer.
2. Do different sound vibrations have the same effect on earthworms? Select a group of students to put an earthworm in a small beaker, set the beaker in a tray of water, and strike the edge of the tray. Have them make several repetitions with different earthworms.
3. Ask several pupils to test whether an animal's size influences its reaction to vibration.
4. Provide students with reference books or library facilities and assign the following questions:
   a. How does an insect's nervous system prepare him to meet emergencies?
   b. How is an insect's sense of touch highly developed?
   c. Do all insects have highly developed vision? Explain your answer.
   d. What insects possess hearing organs on their body? How do they use them?

Demonstrations And Experiments

1. Describe to the class how scientists condition certain animals. Demonstrate this principle by conditioning an earthworm to turn left in a T-maze. Place copper electrodes attached to wires connected to a 1 1/2 volt battery in the right arm of the maze. The

worm, after receiving shocks for a wrong turn, soon "learns" to stay left. If you have sufficient materials, allow students to do this experiment.

2. Hydra, a small fresh-water polyp, shows interesting behavioral characteristics which can be shown with an overhead projector. Half fill a glass tray with distilled water. Place a hydra into the tray and set it over the projector. Use a different hydra for each of the following demonstrations:
   a. Touch various parts of the hydra's body with a dissecting needle.
   b. Send a weak electrical shock through the water.
   c. Add a drop of vinegar to the water.
   d. Set a vibrating tuning fork next to the tray.
   e. Shine a bright light over the tray.
   f. Slowly increase or decrease the temperature of the water.

3. Ask students to fill a 250 ml beaker with equal amounts of leaves, sand, garden soil, and dry paper. Tell them to pour the sand in first, followed by garden soil, leaves, and dry paper. Then have them measure and record the temperature of each level. Give them several earthworms to drop in the beaker. After several hours, ask pupils to remove the material from each level and find their worms. Have them report their findings.

4. How long will sowbugs remain in a curled up position? Ask students to devise a way to find out. Suggest they use a variety of stimuli.

5. Which animal is more sensitive to light—a planaria or fruit fly? Have students put this question to a test.

DISPROVING THE THEORY OF SPONTANEOUS GENERATION

Tell students that for centuries many people believed life originated from non-living matter. They felt worms popped out of rain barrels, maggots (fly larvae) sprang from meat, and geese "grew" from barnacles. This was known as spontaneous generation.

Relate the story of Jean Baptiste van Helmont, Belgium physician, who believed the following recipe produced mice: Place a dirty, sweaty shirt over some wheat kernels in a dark place for 21 days. Mention that Francesco Redi, seventeenth century scientist, demonstrated that maggots and flies are produced from meat only if living flies have laid eggs on the material. Louis Pasteur, nineteenth century French scientist, demonstrated bacteria were living organisms and came from other living organisms.

ACTIVITY 12

Give each group of students 3 - 250 ml Erlenmeyer flasks (or bottles). Have students place a small piece of spoiled meat in each container. Tell them to leave the first container open, tie cheese cloth over the mouth of the second, and stopper the third container. Then place these containers where flies can find them. Students should check each flask daily for 10 days and keep an accurate record of their observations.

ACTIVITY 13

Make beef broth by adding a bouillon cube to warm water. Slowly stir the broth until the cube dissolves. How much you make depends on class size. Let broth stand for 2 days. Sterilize broth in a pressure cooker for 15 minutes at 15 pounds of pressure. Have students fill 2 sterilized glass Petri dishes one-fourth full of broth. Tell them to seal, with tape, the lid onto one dish and leave the remaining dish open. They should make daily observations for the next 10 days. Encourage students to use microscopes or magnifying glasses to help them with their investigations.

Suggested Problems And Questions

1. Ask students the following questions:
    a. Do these experiments prove that "Life comes from life"?
    b. What experimental evidence can you give against spontaneous generation?
    c. How might living organisms enter a sterilized, sealed container?
2. Here's a springtime demonstration: Have students make a yeast cell solution, set it aside, and make daily observations. Fruit flies gather around the fermenting solution. Ask pupils to set up an experiment to prove fruit flies were not spontaneously regenerated.

Demonstrations And Experiments

1. Students can make culture plates from agar medium by following the directions on the bottle label. Have them determine which of the following conditions support life:
    a. A sterilized, open container
    b. An unsterilized, open container
    c. A sterilized, closed container
    d. An unsterilized, closed container

2. Demonstrate the idea of spontaneous action. Place a few thymol crystals on the bottom of a Petri dish. Slowly heat the crystals until they melt. Put the dish on an overhead projector, turn off the lights, and add a thymol seed crystal to the melt. Students will be impressed by the rapid crystal growth.

   Caution: Thymol mildly irritates the eyes and skin. Keep fumes away from face and hands.

   Remind students this is a chemical activity, not life.

HOW DO LIVING ORGANISMS DUPLICATE THEMSELVES?

Tell students life continues through the process of reproduction. Each kind of living organism has the ability to produce an exact replica of itself. Organisms are capable, through metabolism, of using materials within their bodies to conceive new life.

Use available films, film strips, models, or transparencies to illustrate the different methods of reproduction—fission, budding, spore formation, vegetative process, regeneration, pollination, and sexual reproduction. Available materials, allotted time, student interest, and district policy determine which area to stress.

ACTIVITY 14

Begin a discussion of mitosis—cellular division—by asking students a puzzling question: What multiplies by dividing? You'll receive some interesting answers.

Tell students single-celled organisms generally reproduce by mitotic division. This process involves an equal division of nuclear chromatin material from one parent cell into two like daughter cells.

Use charts or board diagrams to explain each mitotic stage—prophase, metaphase, anaphase, and telophase. Relate how chromatin material develops into chromosomes which change shape, divide into equal halves, move toward respective poles, and revert to the netlike pattern of the metabolic cell. Explain mitosis occurs during embryonic development, body tissue repair, and growth.

Fill the bottom of a Petri dish with corn oil. Add iron filings (make a circle approximately one inch in diameter) to the center of the dish. Place a small bar magnet on an overhead projector, lay the dish on the magnet, and move it slowly left to right across the magnet. The iron filings rearrange themselves; some migrate near each pole simulating chromosome division during anaphase (Figure 5).

Let your students duplicate the various stages of mitosis at their desks or lab tables. Give them Petri dishes, magnets, corn oil, iron

Figure 5

filings, and glass probes to stir the filings. Ask pupils to solve the following problem: Use one (or two) magnets to duplicate interphase, anaphase, and telophase. Permit students to continue making investigations on their own.

Again remind students this is simulation, not life.

Suggested Problems And Questions

1. Have students report on the function of DNA and RNA in the cell.
2. Assign a group of students to report on cancer cell development.
3. For extra credit, give students the following questions:
   a. What is amitosis? How does it occur?
   b. How does meiosis differ from mitosis?
   c. What is the function of spindle fibers and centrioles in mitosis?

ACTIVITY 15

Tell students Rhizopus nigricans, bread mold, illustrates reproduction through sporulation. As the mold grows, threadlike mycelium grow upward and develop round bodies or sporangia. These sporangia carry spores which, under favorable conditions, burst open and germinate into a new plant.

Part A

Have students make bread mold by dampening a piece of bread, placing it in a Petri dish or beaker, and leaving it exposed in a warm, dark place for several days. A layer of small, stringy filaments soon coat the bread.

Ask students to examine some bread mold under a microscope. Tell them to break open a sporangium with a needle, make a slide, and examine the contents.

Part B

Have pupils collect as many bread mold spores as possible. Tell them to transfer these spores to Petri dishes arranged with the following conditions:

1. Moist cracker crumbs, place in dark area.
2. Dry bread pieces, place in dark area.
3. Cheese slice, expose to light.
4. Moist bread pieces, heat for 15 minutes in a water bath, expose to light.

Tell students to use several kinds of bread. Those with preservatives don't work well for this experiment.

Go over the results in class.

Suggested Problems And Questions

1. Do bread mold spores germinate on other foods? Have students sprinkle bread mold spores on a bacon strip, apple slice, and a layer of tomato paste. Tell them to keep these materials in a warm, dark area for several days.
2. Are mold spores hard to kill? Ask students to collect various molds. Heat them for 20 minutes in a water bath or pressure cooker. Transfer to a moist medium, and set aside in a warm, dark area for several days. Have them report their findings.
3. How do molds respond to extreme cold? Tell students to duplicate the above problem only freeze the mold for several hours before transfering them.

Demonstrations And Experiments

1. Have pupils dry out some bread mold and sprinkle over different moist food products. Ask them to report on which products, if any, the mold germinates. Remind students to keep mold in a dark, warm area.
2. Demonstrate vegetative reproduction. Cut pieces of geranium or rose plant stems; plant in topsoil or sand; and encourage students to make daily observations.
3. Mention regeneration, the renewal of a lost body part, occurs in certain animals, i.e., starfish and planaria. Give students Petri dishes, pond water, droppers, razor blades, and planaria worms. Tell them to lay a worm in the dish, drench with pond water, and cut the worm into several pieces—quarters, thirds, halves, etc.

Remind pupils to feed the planaria (pieces of liver, or hard boiled egg). Keep covered, and add water to prevent evaporation. Have students make daily observations for three weeks.

4. Allow students to investigate mitosis in an onion or horse bean root tip. Give them these guidelines: Place a piece of root tip on a glass slide, lay a cover slip over the root tip and carefully press downward to separate the tissue, add a basic stain (methyl green, aceto-carmine, aceto-orcein). Have pupils view under high and low power and sketch the various stages of cell divisions.

5. Discuss how yeast cells reproduce by budding, a process whereby a large cell divides repeatedly, forming new cells. Have students prepare a yeast solution, make several microscopic slides, and search for budding cells.

6. What is germination? Give students some ripe pollen grains. Have them make slides and observe under low power and sketch what they see.

    Have pupils make a dilute sugar solution and fill the bottom of a Petri dish. Give them pollen grains to spread over the surface of the water. Tell them to cover their dishes and set aside for 2 days. Ask students to make additional microscopic slides and sketch the germinating seeds.

7. Have pupils sprinkle pollen grains over Petri dishes containing the following ingredients:
    a. A ten percent molasses solution. Cover and keep in a warm area.
    b. A ten percent vinegar solution. Keep uncovered in a refrigerator.
    c. A ten percent salt solution. Cover and keep in a cool place.
    d. A ten percent fruit juice solution. Keep uncovered at room temperature.

    Ask students these questions:
    1. What conditions are necessary for germination?
    2. Why do some pollen grains fail to germinate?
    3. Which container offered the most suitable environment?

CAN UFO'S BE REPRODUCED IN THE CLASSROOM?

Close the Life Unit with a discussion on UFO (Unidentified Flying Object) sightings and the possibility of life on other planets. Tell students a suitable temperature, sufficient oxygen, and quality soil enable living organisms to thrive on earth. No other planet in our solar system enjoys these earthly qualities. Some scientists, however, feel Mars, with its changing landscape coloration, might sustain simple plant life. They feel lichen, a plant composed of alga and fungus, may be such an organism.

ACTIVITY 16

Inform students lichen lives on rocks, tree trunks, posts, and buildings in all kinds of weather. Emphasize algae and fungi form a real partnership. Algal cells share food with the fungi which is unable to produce its own. In return, the fungi collects and retains the water which algae require for the manufacture of food.

Give students several lichen samples. Have them pull a lichen apart, examine thin portions under a microscope, and locate the fungus—threadlike structures—beneath the algae.

Ask pupils to set up a series of experiments to see which of the following conditions are the most destructive:

1. Extreme cold
2. Extreme heat
3. Very moist or dry conditions
4. The effect of different chemicals (acids, bases, etc.).
5. The effect of different concentrated gases (ammonia, carbon dioxide, oxygen, hydrogen, and methane).

ACTIVITY 17

Tell students a 1937 radio dramatization of a Martian invasion created panic throughout the country. Many people believed we were under actual attack.

Give your students the opportunity to witness—even in-vestigate—a simulated invasion. Fill several small capsules or test tubes with 0.5g of sawdust, 0.5g of sand and 0.5g of dry powdered yeast. Mix the ingredients by shaking each tube. Prepare several liters of ten percent sugar solution. Do not reveal the ingredients to your students.

Ask students to do the following:

Part A

1. Place a light source in the middle of the laboratory table.
2. Fill a 250 ml beaker half full of solution. Place the beaker next to the light, keeping the beaker between you and the light.

3. Open the invasion capsule, turn on the light source, turn off the room lights, and pour the ingredients over the surface of the solution.
4. Write down your observations. Students will see some particles (sawdust) float, others quickly sink (sand), and several slowly fall to the bottom (yeast). Yeast cells give off a cloudy trail and "explode" upon hitting the bottom.
5. After the reaction, place the beaker aside and let stand overnight.

Part B

1. With a dropper, take a sample of solution from the beaker, make a slide, observe under low power, and sketch what you see.
2. Take samples from different levels—top, middle, bottom—and observe under low power.
3. Answer the following questions:
   a. Explain how each capsule ingredient differs.
   b. Which particles brought the "visitors from outer space"? List the evidence to support your answer.
   c. What life processes do you observe?
   d. Can you identify the "visitors"?

ELEVEN SUGGESTIONS FOR FURTHER STUDY

If students wish to experiment further, offer them any of the following items:

1. Dissect a large flower. Pull off the petals. Locate the stamens and pollen. Remove the stamens from the pistil. Open the pistil and examine the ovules. Give the function for all of these structures.
2. Make a report on how insects pollinate flowers.
3. Can mitotic division be studied in a radish root? Set up an experiment to find out.
4. On a piece of moist cloth evenly distribute bean seeds. Roll up carefully, do not fold tightly. Moisten cloth roll, place in a jar or beaker, and cover lightly. Examine every other day for eight days. Keep accurate record of your findings.
5. Boil several bean seeds for 20 minutes. Repeat the above process, only include an equal number of boiled seeds with fresh bean seeds. Do any of the boiled seeds germinate?
6. Can bean seeds grow in any medium? Place an equal number of bean seeds in trays containing different mediums—crushed gypsum (chalk or plaster of Paris), crushed charcoal, and garden

topsoil. Expose each tray to the same temperature, amount of light, and moisture. Report all findings.

7. Set up a display of life processes. Include examples of food-getting, digestion, absorption, assimilation, movement, sensitivity, reproduction, respiration, circulation, and excretion.
8. Make a report on the carbon dioxide-oxygen cycle and the nitrogen cycle.
9. Set up a demonstration of diffusion and osmosis.
10. Research and report on the Heterotrop hypothesis.
11. What is the difference between asexual and sexual reproduction?

chapter two

19 ACTIVITIES THAT
TEACH ASTRONOMY

OVERVIEW

In this unit the student examines the sun's energy, the sun's family of planets and the sun itself.

Pupils explore a mystery planet, make stony and metal "meteorites," and create "tektites" from molten glass.

Many activities lead students into a study of stars and constellations. They experience building a reflecting telescope from an inexpensive shaving mirror and cardboard tube.

The unit closes with pupils making rockets from paper matches.

It's a rare person who can walk by a telescope and not glance toward the heavens. People of all ages enjoy piercing the sky hoping to free a distant star, glowing comet, or speedy meteor.

Ancient sailors depended on stars to guide their ships. A clear night assured a safe arrival; a foggy evening sent many ships into the jaws of coral reefs.

Groups of stars, called constellations, were named after mythical figures—bears, dragons, scorpions. The Egyptians, Greeks, and Romans believed gods ruled the stars and planets.

Why study Astronomy? The knowledge we receive from the universe around us helps us to enjoy and appreciate living. This knowledge has changed our way of living, replacing ignorance and superstition.

WHAT ARE SOME WAYS TO STUDY LIGHT ENERGY?

Open this unit with a discussion of the sun. Tell students primitive people considered the sun as the source of all life. Legends extend as far as space itself. The Egyptians believed Atum-Ra was the ruler of creation. The Suma Indians of Central American thought Papan, an early wanderer, rose toward the heavens and became the sun.

Remind students superstition and fear guided the lives of early people. Once these fantasies gave way to knowledge, the heavens opened up and spilled forth new discoveries.

ACTIVITY 18

Tell students the sun provides both visible and invisible radiant energy and scientists estimate from 35 to 43 percent of solar energy reflects back into space. Ask pupils to find out what surfaces absorb the most energy. Have them set a tray of leaves, moist soil, dry soil, and fresh water the same distance from a heat lamp. Remind them to keep the surface level of each tray the same.

Have pupils take the temperature of each tray at 2 minute intervals for 16 minutes. Then graph the results.

Suggested Problems And Questions

1. Ask students which surface absorbs the most heat? The least heat? Have them account for their answers. Dark surfaces absorb more heat than shiny surfaces.
2. Have pupils list the variables which might effect experimental results. Answers will vary.
3. Tell students to set up a test to find out which of the following surfaces have the best absorbing power: Fresh water, silty water, salt water, and aluminum foil. Silty, dirty water should show a greater absorbing power.
4. Have students test the absorbing power of moist soil set at different angles to the light source. Which angle reflects the least amount of radiant energy? Answers will vary. *Hint:* The angle of incidence equals the angle of reflection.
5. Does a dark or light surface absorb the most heat? Let students investigate this problem on their own.

ACTIVITY 19

Explain that not all of the world's energy remains above the earth's crust. For millions of years the earth has stored oil, coal, and sea water in her crustal vault. These fossil fuels bulge with the sun's trapped energy and escape when used by man.

Give students the following materials: burner, large test tube with one-hole stopper, ring stand and clamp, powdered coal (bituminous), 250 ml Erlenmeyer flask with two-hole stopper, large beaker (500 ml) with water, a glass jet tube, and a long glass tubing set at a right angle (Figure 6).

Have pupils set up apparatus as shown in Figure 6. Tell them to half fill the test tube with coal powder, heat by moving the flame of the burner back and forth across the bottom of the test tube. Gas and chemicals will collect in the beaker.

Suggested Problems And Questions

1. Ask students to perform the following activities:
   a. Hold a lighted splint over the jet tube. Record what happens.
   b. Describe the odor of the gas.
   c. What chemicals are left in the beaker?
   d. How does this experiment relate to our air pollution problem?
2. Have pupils report on how coal or oil is formed in the earth.

ACTIVITY 20

How To Make A Solar Cooker

Emphasize coal, fuel oil, peat, lignite, and natural gas are fossil fuels. These fuels are non-renewable. Some countries lack these fuels or are running very low. These conditions add to the energy crisis problem.

Some countries with a scarce fuel supply use solar cookers to provide the needed heat. Have students build and test a miniature solar cooker. Give them the following directions:

1. Cut out a circle with an 8 inch diameter from plain white paper.
2. Make a paper funnel by folding the circle in half twice (Figure 7). Tape down the loose edges.
3. Spread the sides of the paper apart to form a cone shape. Set in a small beaker (50-100 ml).
4. Line the inside walls with aluminum foil.
5. Add a small lump of clay to the funnel bottom.
6. Insert a piece of glass tubing (3 inches) into a small one-hole rubber stopper. Push the glass tubing into the clay. Make sure the tubing stays in an upright position.
7. Darken the outside of a 20 ml beaker. Do this by holding the beaker over a lighted candle.
8. Put 10 ml of water in the beaker and set it on the rubber stopper.
9. Adjust your ring stand, clamp, and light source to shine directly over the beaker (Figure 7).
10. Take temperature readings every 2 minutes for 10 minutes. Graph the results.
11. Repeat the above procedure using a plain white funnel and clean, unsooted beaker. Graph the results.

Figure 7

Figure 6

Suggested Problems And Questions

1. Have pupils answer the following questions:
   A. Why is it necessary to cover the inside funnel walls with foil? To increase reflecting power.
   b. What is the purpose of coating the water-containing beaker? To increase heat absorption.
   c. What did this experiment show? Reflected light will be absorbed by the carbon-coated beaker.
2. If you have time, ask students to prove or disprove the following statements:
   a. The closer the light source, the more heat is absorbed.
   b. The larger the solar heater, the more heat is absorbed.
   c. The steeper the light source angle, the more heat is absorbed.
   d. A salt solution absorbs more heat than a fresh solution.
   e. A paper funnel lined with microscopic glass slides will reflect more heat than one with aluminum siding.
3. Light travels at approximately 186,000 miles per second. Have students figure how long it will take light to travel from the sun to each of the planets listed on the following chart:

Planet	Distance From Sun (Million Miles)	Traveling Time
Venus	67	
Mars	142	
Saturn	886	
Uranus	1,782	

ACTIVITY 21

Discuss the source of the sun's energy. Mention most scientists believe hydrogen atoms in the interior of the sun change into helium atoms. This conversion process involves a small loss in mass. Einstein believed this matter is converted into energy and huge amounts of energy came from small quantities of matter. During this change, some matter is converted into light and heat energy. For this reason, the sun is gradually losing weight.

Tell students to do the following:

1. Weigh a 50. ml beaker.
2. Weigh 20 cc of dilute hydrochloric acid.
3. Weigh a 3 inch strip of magnesium ribbon.
4. Find the total weight.
5. Roll up the magnesium ribbon and drop it into the acid solution.
6. After the reaction, weigh the beaker contents.
7. Report your findings. How does this experiment relate to the discussion on energy conversion?

Describe the various parts which make up the sun—photosphere, chromosphere, corona, solar flares, prominences, and sunspots. Use diagrams, charts, and a slide or film strip projector to supplement your discussion.

ACTIVITY 22

Tell students sunspots appear as storms of hot gas twisting into the sun's atmosphere. Some scientists believe magnetic disturbances within the sun forces huge masses of gas through the sun's surface and into the atmosphere. They vary in size, rotate, increase and decrease in regular cycles, and come and go.

Have students fill the bottom of a Petri dish with hydrogen peroxide. Then place a quarter-size pile of manganese dioxide into the center of the dish. Tell them the manganese dioxide represents the sun. Have pupils repeat this experiment two or three times before answering the following questions:

1. The corona is a fringe of glowing gases encircling the sun. How does this experiment demonstrate the sun's corona?
2. A solar flare is a tremendous outburst of energy from the sun. How does this experiment simulate solar flares?
3. Do "sunspots" appear, disappear, and reappear again? How does this phenomena occur during the reaction?

ACTIVITY 23

How To Make Your Own "Aurora Borealis"

Mention how people who live in the extreme north or south latitudes frequently see a beautiful blend of colored light flowing through the sky. This colorful radiation constantly changes shape. It ranges from no color at all to violet, yellow, orange, lavender, and green.

People who live in the north call these northern lights aurora borealis. People who live in the south refer to them as the southern lights or Australis borealis.

Tell students these lights seem to be associated with solar flare activity. When intense solar flares occur, magnificent aurora displays soon follow.

What causes the auroras? Some scientists think that electrically charged particles from the sun interact with the earth's ionosphere.

Hang different colored paper around the room. Provide students with number four one hole rubber stoppers and four inch pieces of glass tubing, one quarter inch diameter. Have them insert the tubing into the stopper, bring the stopper next to the eye, look through the glass tubing, and move it slowly in different directions across the colored paper, e.g., up, down, sideways, diagonally and in circles. Students will see a beautiful display of simulated auroras changing shape and color. Ask them to record their observations.

Let them investigate further with different lengths and diameters of glass tubing.

Demonstrations And Experiments

1. Fill the bottom of a Petri dish with a clear oil, add a small pile of iron filings, and place a bar magnet near the bottom of the dish. Move one end of the magnet back and forth across the bottom. The iron filings represent sunspots. Do this demonstration on an overhead projector.
2. A solar eclipse occurs when the moon lines up between the sun and the earth. The moon's shadow strikes the earth's surface.

    Demonstrate a solar eclipse. Suspend a large Sytrofoam ball (earth) from a ring stand. Darken the room, shine a flashlight (sun) against the earth. Ask a student volunteer to pass a small Sytrofoam ball (moon) suspended from a string between the sun and earth. The moon's shadow falls on the earth.
3. The sun changes location in the sky throughout the day. Students can experiment with these changing positions to determine time

by using a sundial. The sundial, to be parallel to the earth's axis, should point to true north-south and be set at the latitude angle for the student's region.

Students can construct a simple sundial by setting a pencil into a lump of modeling clay in the center of a paper circle (filter paper) with a scale designed to mark the shadow every hour.

4. The photosphere, or sun's "surface" is characterized by markings resembling rice grains. Astronomers say each "grain" is about 700 miles wide and about 200 miles deep. They may be caused by hot gases rising to the surface.

Demonstrate by pouring a layer of Rice Krispies over a Petri dish containing water. Set the dish on an overhead projector.

5. Sunspots may be observed by placing a piece of white paper or cardboard behind the eyepiece of a telescope which is pointed toward the sun. Adjust the paper until a sharp image appears. If you fail to see sunspots, try again. They invariably come and go at different intervals. *Caution:* Under no circumstances look directly into the telescope. The sun's rays can cause permanent injury to the eyes.

6. Simulate an eclipse with an overhead projector. Trace the edge of a large paper filter disk on the projector glass with a grease pencil (sun). Slowly move the filter dish (moon) across the outline. As the black circle (moon) crosses the circle of light (sun), an eclipse will appear.

EXAMINING THE NINE PLANETS

Most students have had earlier work in astronomy—especially material covering the nine planets in our solar system. Teachers use various methods of presenting the proper sequence of planets from the sun. Here is one scheme: If we let the letters in the sequence, MVEMJSUNP, be the first letter of each word in a telegram, we get the following: Many veterans entering Mason Junction Saturday. Urge nice parade (Mercury, Venus, Earth, Mars, Jupiter, Saturn, Uranus, Neptune, and Pluto). Few students are likely to forget this patriotic message.

ACTIVITY 24

Read the following story to your class and ask students to write down or list the main points emphasized for each planet.

Mother Sun had 9 children. She named them Mercury, Venus,

Earth, Mars, Jupiter, Saturn, Uranus, Neptune, and Pluto. She clucked over them like a proud mother hen.

Whenever there was a celestial gathering, she made sure everyone heard her story. "Mercury," she'd begin, "is my *closest*. A very *warm* child. Venus, my beautiful, shy daughter, hides from me under a *cloud-shrouded* cape. Earth respects me, but allows herself to be *abused* by her inhabitants. Mars, the story teller, boasts that he sends terrible *beasts* and *monsters* searching throughout the universe for unsuspecting victims. Jupiter, my *largest* offspring, loves to play tricks. His favorite is moving a large, *red spot* across his chest. He drives people crazy. Saturn, a proud youngster, moves through the sky surrounded by beautiful *rings*. A truly angelic child. Uranus, Neptune and Pluto live *very far away*. They *seldom come around*. They wish to be alone and *away from* their mother's warmth."

List the pupil responses on the board and discuss them.

ACTIVITY 25

Ask pupils why it takes Pluto longer to circle the sun than Mercury. Answers will vary. However, most students agree a planet's distance from the sun largely determines this time.

Part A

The chart below lists the planets, their approximate distances from the sun, and the approximate time it takes each planet to orbit the sun. Using the scale 1 centimeter equals 36 million miles, have students estimate each planet's distance from the sun (in centimeters). Answers are provided on chart.

Planet	Approximate Distance From Sun In Miles (million)	Approximate Time To Orbit Sun	Approximate Scale (cm)
Mercury	36	88 days	(1.0)
Venus	67	225 days	(2.0)
Earth	93	365 1/4 days	(2.5)
Mars	142	687 days	(4.0)
Jupiter	489	12 years	(13.5)
Saturn	890	30 years	(25)
Uranus	1,785	84 years	(50)
Neptune	2,795	165 years	(78)
Pluto	3,700	280 years	(103)

Part B

This activity emphasizes one point: The further a planet is from the sun, the longer will be its period of rotation (no allowance is made for elliptical orbits).

Ask pupils to cut 9 pieces of string or thread the centimeter length of each planet's distance from the sun calculated in Part A. Form a loop on one end of the string, place a pencil through the loop, tie a small swivel lead weight (rubber stopper or heavy washer) to the other end. Push the weight hard enough to swing around the pencil several times. Make sure the distance between the pencil point and weight attachment agrees with the previous calculations.

Tell students to time each orbit with a stop watch. Encourage them to clock at least 5 swings per planet and find the average time. Have them graph the results by placing *planet* on the vertical axis; *the average time per planet* on the horizontal axis.

Suggested Problems And Questions

1. Ask students to write a report on their experimental findings, including their graph results.
2. Have students pick one of the following topics and report to the class.
   a. The Ptolemaic System of planetary motion (after Claudius Ptolemy).
   b. The Tychonic System (after Tycho Brahe).
   c. The Laws of Kepler (after Johannes Kepler).
   d. Newton's Law of Universal Gravitation.
3. Encourage a group of students to demonstrate Newton's Laws of Motion: Inertia, relationship between force and mass, and examples of force.

ACTIVITY 26

Plans For Building Your Own Planet

Tell students they will investigate a newly discovered planet—Planet Enigma. The following are the steps necessary for constructing this celestial body:

1. Take a 1,000 ml beaker (or any medium-size container), fill it with soil, add water, mix, and pour the muddy batch into a Pyrex (optional) tray. Be sure to make enough to cover the entire tray (approximately one inch thick).
2. Strengthen the mixture by adding small boulders, pebbles, or sand.

3. Mold valleys, canyons, hills, and mountain peaks.
4. Insert 3 small crucibles into the crust. These liquid-filled containers make excellent lakes or seas. Pick any of these aqueous recipes:
   a. Red dye in a 5, 10, or 20 percent salt or sugar solution.
   b. A yeast cell suspension (prepared by adding a half-teaspoonful of powdered yeast to one bottle of sugar solution).
   c. A dilute ammonia or alcohol bath.
   d. A saturated potassium dichromate or copper sulfate solution.
   e. Blue dye with plain water.
   f. A solution containing one-celled organisms (amoebae, paramecia, etc.)
5. Sprinkle iron filings, charcoal or lead powder, sulfur, boiling chips, copper sulfate crystals, and aluminum powder over the surface.
6. Poke small twigs, weeds, root fibers, or grass blades into the soil.
7. Allow the planet to dry at room temperature for two or three days. The drying process produces cracks resembling fault lines (Figure 8).

Figure 8

Give each student a sample observation chart (Figure 9). This should be used only as a guide. Encourage pupils to make their own interpretations and recordings as they see fit. Provide the following equipment:

1. Test tubes and tongs.
2. Alcohol or Bunsen burners.
3. Forceps and/or tweezers.
4. Scoops.
5. Compound microscopes and slides.
6. Dissecting microscopes.
7. Magnifying glasses.
8. Magnets.
9. Litmus paper.
10. Cobalt chloride paper (test the presence of water.)
11. Beakers (50-100 ml).
12. Baby food jars, small test tubes, or beakers for collecting and separating materials.
13. Mortars and pestles.
14. Thermometers (to take different soil level and basin temperatures.)

Sample Observation Chart

	Basin No. 1	Basin No. 2	Basin No. 3	Geologic Structures	Vegetation	Living Organisms	Other
Composition							
Shape							
Color							
Location (N,E,S,W)							
Temperature							
Stage of Development							
General Condition							
Unique Features							
Additional Comments							
Sketch or Drawing							

Figure 9

RECIPES FOR MAKING "TEKTITES" AND "METEORITES"

ACTIVITY 27

Open a discussion on "Space Garbage." Pupils generally find tektites, meteors, meteorites, and unidentified flying objects interesting.

Begin by introducing tektites. Stress the following points:

1. Some scientists believe tektites are meteorite pieces, or the results of impacts on earth by meteorites, asteroids, or comets.
2. They have a natural glass consistency (pass around synthetic glass samples or the real thing, if you have them).
3. They range in weight from under a gram, to approximately eight kilograms.
4. Tektites may be black, green or yellow; they may have a round, elliptical, teardrop, dumbbell, or disk and button shape.

After discussion, let pupils make their own tektites. Provide the following equipment.

1. Bunsen burner.
2. Different sized glass tubing (small diameters work best.)
3. One-holed rubber stoppers (these make excellent glass holders.)
4. Tweezers or forceps.
5. Paper toweling.
6. Collecting tube and cap.

Keep directions brief and simple. Allow students free rein in creating their products. Give them these guidelines:

1. Never touch hot glass with your hands. Place a rubber stopper on each end of the glass tubing. These holders make it easy to rotate the glass tubing over the flame.
2. Adjust the burner to produce a fairly hot flame. Slowly turn the glass until it begins to soften.
3. Don't attempt to pry the glass tubing apart. Allow it to bend with the force of gravity. Use the tweezers to snap off or shape molten end pieces.
4. Have students make iron tektites by sprinkling fine iron filings over a molten piece of glass. The filings give off a shower of sparkles but will readily stick to the glass. Pressing these end pieces against paper toweling produces excellent black tektites.
5. Tell pupils to make their final end products small enough to fit into the collecting tubes. Each student writes his name on a piece

of masking tape, wraps it around the collecting tube, and turns it in at the end of the period. This helps the teacher keep track of them.

6. All students should wear protective safety glasses.

Let students discover for themselves how to push, pull, twist, or turn their glass pieces to get a desired shape, i.e., slowly pulling the tubing apart produces a teardrop design; gently pushing two pieces of glass together, with a slight twist, creates a button and disk shape. (Figure 10).

Hold a contest. Have students make as many tektites as they can within a class period. Award one point per tektite.

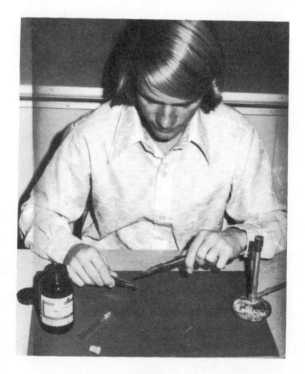

Figure 10

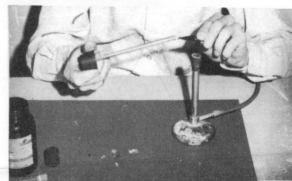

ACTIVITY 28

Relate how meteorites, meteors which reach the earth, dimple the earth's surface. They level forests, punch holes in roof tops, and carve giant craters in the land. They've been around since the beginning of time.

Some scientists believe meteorites are fragments from the "missing planet" (a planet which some astronomers think once roamed between Mars and Jupiter). Other investigators feel they come from the tails of comets. When meteors approach the earth, the gravitational field, like a powerful magnet, pulls them closer. The friction which builds up in the atmosphere creates intense heat.

A meteorite strikes the earth, scatters, and sends shock waves racing in every direction. The depth of penetration and the crater size depend upon where the meteorite hits the earth (soft or hard terrain), its speed, and the size of the meteorite at impact. The crater is generally larger than the meteorite.

Have students make a clay (or semi-dry mud) pad (5" x 3" x 1"). This will represent the earth's crust. Give them three different size steal bearings—small, medium, and large. Each bearing simulates a meteorite. Students should follow these procedures:

1. Place the clay pad in the center of the laboratory table or on the floor.
2. Stand over the clay pad, hold the bearing at shoulder level between your thumb and index finger. Center the bearing over the pad and release it (Figure 11).
3. Drop each bearing, one at a time, on the clay pad.
4. After the first bearing hits the clay, remove it carefully; then measure (mm) and record the depth and diameter of the crater.
5. Repeat, taking the average of several drops.
6. (Optional) pupils can time the falling speed of the bearing with a stop watch.
7. Have them plot *time* and *bearing size* (diameter) on graph paper.

Give students these questions to answer:

1. Which bearing makes the largest crater? (The heaviest.)
2. Which bearing travels the fastest? The slowest? (The heaviest travels faster; the lightest moves slower.)
3. Are the crater and "meteorite" the same diameter? Explain. (No. Impact forms a larger crater size than meteorite.)
4. What can you say about the crater size and speed of a meteorite? (The faster the meteorite, the larger the crater.)

Figure 11

5. What factors determine the speed of a meteorite? (Air friction, size and density of meteorite, etc.)

ACTIVITY 29

Tell students few meteorite craters are found on earth because most meteorites land in the ocean, burn up in the atmosphere, or erode and weather away before they are discovered. However, scientists recognize two main kinds of meteorites: Iron-nickel (siderites) and stony (aerolites.) The iron-nickel meteorite generally contains 90 percent iron, 8 percent oxygen, 24 percent nickel, 18 percent silicon, 14 percent magnesium, plus additional substances.

Stony meteorites come in two subgroups:

a. Chondrites: An aggregate of rounded grains containing olivine and pyroxene.

b. Achondrites: A combination of plagioclase feldspar, pyroxene, calcium, iron, magnesium, and silica.

Part A

Have students make a "stony meteorite" from the following recipe:*

> 5.6 g sand (silicon dioxide)
> 1.0 g salt (sodium chloride)

1.0 g iron powder or filings
0.4 g magnesium sulfate, crushed dolomitic
   limestone, or talc (magnesium products)
<u>1.0 g</u> plasticine clay (adhering agent)
9.0 g Total

Part B

Have students make a "metallic meteorite" from the following recipe:*

7.0 g iron powder or filings
1.0 g nickelous sulfate
1.0 g plasticine clay (adhering agent)
0.5 g aluminum powder (provides metallic
   sheen)
0.5 g charcoal powder (provides dark
<u>   background)</u>
10.0 g Total

Part C

Make a batch of "mystery meteorites." Use this recipe:

 50 g crushed pyroxene, olivine, or sand
100 g crush plagioclase feldspar or talc
 25 g iron powder or filings
<u> 25 g calcium carbonate (crushed calcite)</u>
200 g Total

Add enough clay to hold the mixture together. Mold approximately twenty "meteorites"—a la—the stony variety. Pass them out and present this problem to the pupils: You're hiking in the country and accidentally stumble across a strange looking rock. It's different from anything you've ever seen before. You suspect, however, it might be a meteorite, but what kind? Rip it apart and find out. Use a microscope to aid your investigation.

Assign the following questions:

1. Meteors travel around the sun at speeds up to 25 miles per second. How many miles per hour is this? 90,000 mph.
2. Why haven't more meteorite craters been discovered? Many land in isolated areas or oceans, or craters erode away.

---

* These recipes will make products approximately one inch in diameter. If a larger specimen is desired, double or triple recipe ingredients.

3. What causes meteors to burn up in the earth's atmosphere? The friction which builds up between meteor and air.

ACTIVITY 30

Finish this unit by mentioning how man's imagination extends as far as space itself. In 1834, a scientist examined a meteorite which fell in France. He claimed that living particles zoomed down to earth on the backs of these "shooting stars." Other investigators report similar finds in different parts of the world. Some claim bacteria reach earth via meteorites. Does this mean foreigners from outer space are visiting us? Critics believe these bacterial hitchhikers have been here all the time. They feel meteorites either collect bacteria in the earth's atmosphere or pick them up upon crashing into the earth's crust.

Have students collect various soil samples from different locations: lawn, backyard, dirt pile, etc. Tell them to do the following:

1. Place soil in a Petri dish.
2. Heat a steel bearing (from previous experiment) with a Bunsen burner.
3. Remove from the flame. Allow bearing to cool. Then place under a microscope. Describe its appearance.
4. Heat the bearing again for several minutes.
5. Remove from the flame and drop it into the Petri dish.
6. Wait approximately 30 seconds before removing the bearing from the dish.
7. Make a second microscopic examination. Describe what you see.

Assign the following questions:

1. How do the first and second observations compare? Soil sample collects on second bearing.
2. What does this experiment suggest? Meteorites pick up debris upon hitting the ground.

Demonstrations And Experiments

1. Meteors which produce large, bright lights are called fireballs. Demonstrate a "fireball." Slowly heat approximately one inch of potassium chlorate in a large Pyrex test tube supported by a clamp and ring stand. When the substance liquefies, insert (using tongs) a long, thin piece of wood into the test tube. Make sure to submerge the stick into the molten solution. The wood self-ignites

and sends brilliant flames shooting out of the test tube. Darken the room for full effect. Be sure students keep a safe distance away.

2. Demonstrate various moon phases by darkening the room, turning on a light source (100-150 watt lamp), holding a light-colored beach ball (moon) at arms length while facing away from the light. The light which falls on the ball produces a full moon effect. Turn and face the light. The ball will be difficult to see. This represents a new moon phase. Crescents can be demonstrated by slowly turning in a complete circle while holding the ball at arms length.

3. Briefly explain how the sun, earth, and moon line up to produce a solar, lunar, and annular eclipse. Pass out various size Styrofoam balls, clay, and pipe cleaners. Ask students to make models illustrating each phenomenon.

4. Ask students to construct a model of the solar system. Provide clay, cotton, filter floss, paper, wire, thread, pipe cleaners, cement, etc., for building materials.

MAKE YOUR OWN STAR CHART

Remind students that the ancients named constellations after animals and people. Show them diagrams, transparencies, or filmstrips of well-known constellations: Big Dipper; The Great Bear; Orion, The Hunter; Leo, The Lion; Draco, The Dragon; and several others. Stress imagination played a large part in naming these star groups. Mention these stellar maps are useful in locating individual stars.

ACTIVITY 31

Part A.

Have pupils make their own star charts. Give them black (blue or brown) construction paper, small stiff paint brushes, and paint. Be sure they wear laboratory coats and cover work areas with newspaper.

Give students the following directions:

1. Lay the construction paper on a flat surface.
2. Dip your paint brush in the paint. Tap the brush against the paint jar to remove excess paint. Note: an old toothbrush works well.
3. Remove the brush from the jar, run your thumb across the bristles as you move the brush back and forth over the paper. Avoid dropping large blobs of paint.

4. Repeat until the entire paper receives an adequate coating.
5. Make several charts using different color combinations.
6. Set charts aside to dry.

Provide pupils with thread, glue, and scissors. Tell them to make Scorpio, Big Dipper, Orion, or any constellation they choose. They can do this by cementing the thread to several "stars" which represent these constellations (Figure 12).

Figure 12

Part B.

Ask students to repeat the procedure in Part A. This time have them use their imagination and make up their own constellations.

ACTIVITY 32

How many stars are there on one chart? Tell students to make a square figure from wire or pipe cleaners approximately one-twelfth the size of a chart previously made in Activity 31. Tell them to toss the figure onto the chart and count every dot within the figure, including those touching the edge of the figure. Then multiply this total by 12. This will give an approximate star count.

ACTIVITY 33

Analyzing A Star's Personality

Mention to students a spectroscope, an instrument that uses a prism to separate light of different wave lengths, is used to analyze chemical elements of distant stars, their distances, masses, speeds, and temperatures. Each chemical element, when it is vaporized and glowing, gives off its own pattern of colors.

Have students pass different compounds over a cool, blue Bunsen burner flame using a nichrome wire loop. Tell them to record the color from each element. Remind them to clean the wire loop after each test by dipping it into a dilute hydrochloric acid solution.

If spectroscopes are available, have pupils examine the colored flame through them. Here is a sample record chart listing various compounds.

Compounds	Color of Flame	Spectroscope Colors or Patterns
Potassium Nitrate		
Copper Nitrate		
Sodium Nitrate		
Strontium Nitrate		
Lithium Nitrate		
Calcium Nitrate		
Barium Nitrate		

Demonstrations And Problems

1. Let students make their own spectral lines. Provide them with pictures of spectral lines. Have them hold microscopic slides over a candle flame until completely coated with carbon. Then give them dissecting needles to scratch out their own patterns.

2. Demonstrate how a prism separates light into its colors. Cut a narrow slip in a piece of cardboard, support it in front of a filmstrip projector. Hold a prism, base up, in front of the cardboard. Darken the room and turn on the projector light. A spectrum will appear on the ceiling.

3. Some constellations in the northern sky appear to move around a circle in the sky, remaining above the horizon. They circle about the north star once every 24 hours. This motion can be

photographed with time exposure film, producing bright, curved star trails.

Students can make their own star trails. Have pupils duplicate polar group constellations (Cassiopeia, Draco, Big Dipper, Little Dipper) on a round piece of paper. Make Polaris the center star. Punch out the center with a pencil, place paper over a record player turntable, and set player speed at 16 rpm. Hold the pencil over each star, one at a time, and trace a star trail path.

4. Provide students with star maps from encyclopedias or reference books. Pass out large pieces of filter paper and have pupils make simplified maps of summer and winter sky constellations. Here are some guidelines:

Summer Constellations	*Winter Constellations*
1. Sagittarius	1. Perseus
2. Draco	2. Cetus
3. Hercules	3. Canis Major
4. Scorpius	4. Columba
5. Bootes	5. Orion

5. Stellar distances are measured in parsecs and light years. A parsec equals approximately 19 trillion miles (19×10^{12} miles) and a light year, or the distance light travels in 1 year, is approximately 186,000 miles per second (6×10^{12} miles/year). One parsec equals about 3.26 light years.

Have pupils fill in the following chart:*

Star	Parsec Distance	Light Years Away
Capella	14	(45.64)
Proxima Centauri	(1.29)	4.2
Alpha Orionis	(92)	300
Gramma Orionis	(66)	215

6. A star's brightness, or magnitude, is expressed as a number. The lower the number, the greater the magnitude. For example, a first magnitude star is 2.5 times brighter than a two magnitude star; A three magnitude is 2.5 times brighter than a two magnitude.

Ask students to list the magnitudes (apparent and absolute) of the 10 brightest stars and tell what factors determine these magnitudes.

---

* Answers given in parentheses

ACTIVITY 34

Pass out star charts, paper, different size nails, scissors and watch glasses. Have pupils make a transparency planetarium by smoking the watch glasses with candle soot and scratching various star groups on them. Also, tell students to cut out paper circles and make constellations by poking holes in the paper with nails. Different size nails can be used to indicate relative magnitude of stars. Allow students the opportunity to show their creations on an overhead projector.

ACTIVITY 35

How To Build A Reflecting Telescope From A Shaving Mirror

Let students build their own reflecting telescopes from inexpensive shaving mirrors. These telescopes are fun to build, light, and easy to store. Shaving mirrors produce some distortion due to surface imperfections, but students will experience the principles of the Newtonian telescope.

Have pupils do the following:

1. Determine the mirror's focal point by reflecting sunlight on white paper until a sharp image appears. Measure the distance (Figure 13).

Figure 13

2. Fit the mirror in a 4 or 5 foot circular packing cardboard tube (check the shipping and receiving department for empty tubes).
3. Set a 1" x 1" plane mirror at a 45° angle on a dowel or still wire. Place the mirror a short distance in front of the focal point. The

45° angle allows light to be reflected through the eyepiece. The eyepiece opening should be centered directly above the plane mirror (Figure 14).

4. Place the eyepiece (magnifying glass—double convex lens) in a cardboard or metal cylinder, insert in the opening directly above the plane mirror, and raise or lower the eyepiece to magnify the image formed by the mirror (Figure 15).

5. Make a support rocker (telescope holder) from wood, or support the telescope on an empty wire reel. Use different colored paint or symbols to decorate the telescope cylinder.

6. Take their telescopes outside and practice focusing in on objects at various distances.

Figure 14

Figure 15

ACTIVITY 36

How To Make Rockets From Paper Matches

Close the astronomy unit by letting your students build paper rockets. A paper rocket consists of a paper match surrounded by aluminum foil. They are easy to build, very inexpensive, and quite challenging.

Begin by giving students paper matches, aluminum foil, paperclips, and toothpicks (Figure 16). Have pupils build their rockets in the following manner:

1. Straighten one of the paper clips. Place it along the upper surface of the match. Lay one end of the clip near the base of the match head. This forms a channel or exhaust system for the missile. A round wooden toothpick may be used in place of a paper clip.

2. Bend the other paper clip to the preferred firing angle—30°, 45°,

50°, etc. This provides an excellent launcher. Pull the inside loop of the clip in an upward direction to the desired angle (Figure 17).

3. Cover the match head and half of the match body with aluminum foil. Be sure to press the foil tightly against the match body. Too many loose ends or open spaces will cause a loss of power.

4. Remove the straight paper clip from under the aluminum foil by pulling it out slowly and carefully.

5. Set the rocket on the launcher. Strike a match, hold it under the paper rocket match head, and wait for the blast off (Figure 18). Caution: Never point the rocket toward another person, and be sure to wear safety glasses!

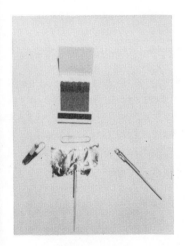

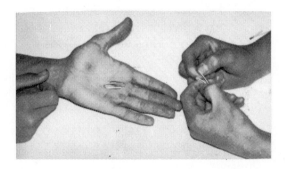

Figure 17

Figure 16

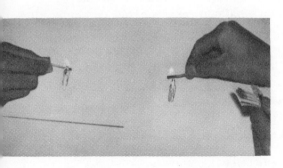

Figure 18

Hold a contest. Encourage pupils to find a partner, and see who can build the best, fastest, or longest flying rocket. Students will probably ask you for suggestions on how to improve their flights. Don't tell them. They'll learn more if they experiment on their own. However, you might mention that any number of factors or variables sneak into scientific experiments and influence the results. Ask your students to test the following variables:

1. Does the amount of foil covering the rocket's body influence its flight?
2. Are all aluminum foil brands alike in weight, thickness, and strength?
3. Do some nose-cone designs (aluminum covering the match head) work better than others?
4. Does it matter how you place the rocket on the launcher—i.e., in front, behind, to the side?
5. Does it make any difference at which angle you set the launcher?
6. Will waxing the launcher's surface increase the rocket's performance?
7. Do wooden matches make better rockets than paper matches?
8. Will wrapping several match heads together produce more thrust, thereby increasing the rocket's travelling distance?

Suggested Problems And Questions

1. Have students research Newton's Laws of Motion and report how they apply to the paper rocket experiment.
2. Ask pupils to set up investigations to determine:
    a. How the amount of force (flaming match) exerted on the match body (rocket) affects its speed.
    b. How the mass (match size) affects the acceleration rate.
3. Suggest that students make balloon rockets. They can determine acceleration rate by blowing up a long balloon, taping a soda straw to it, running a length of string through the straw (length determined by room size), and releasing the balloon. Students should use stopwatches to time several trail runs. Have them graph *time* versus *distance.*

FIVE SUGGESTIONS FOR FURTHER STUDY

1. The Big-Bang Theory suggests the universe was once squeezed together. A violent explosion occurred which threw matter into space. Long after the explosion some of the matter condensed to form stars and planets.

Have students fill small balloons with bits of paper or sawdust, and blow up the balloons until they explode. Ask them to report how this simulates the Big-Bang Theory.

2. Proponents of the dust cloud hypothesis believe huge amounts of dust present in the atmosphere combined to form the solar system. They think radiation pressure and gravity caused dust particles to pull together over a long period of time.

   Tell pupils to cover the bottom of a Petri dish with glycerin and set the dish over a magnet. Then sprinkle iron filings over the surface. Have students answer these questions:
   a. What is the purpose of the glycerin?
   b. Does the magnet simulate radiation pressure or gravity? Explain.
   c. Does this experiment support the dust cloud theory? Why or why not?

3. How do the planets compare in size (diameter) with the sun? Suggest students make models of the sun and each planet from clay. They can use the scale, 1 millimeter equals 10,000 miles. Give them the following information:

Item	Diameter (in miles)	Millimeter Distance
Sun	866,000	86.6
Mercury	3,100	0.31
Venus	7,700	0.77
Earth	8,000	0.80
Mars	4,200	4.2
Jupiter	89,000	8.9
Saturn	71,500	7.1
Uranus	32,000	3.2
Neptune	31,000	3.1
Pluto	3,600	0.36

Optional: If clay is unavailable, have them cut dimensions from construction paper.

4. Give students clay, Styrofoam balls, charcoal powder, wax and glass (to melt), sand, and sculptoring tools. Provide pictures of the moon's surface. Tell pupils to make a model of the moon, including mountains and craters.

5. Encourage a group of pupils to make several constellations by hanging Styrofoam balls from the ceiling.

chapter three

13 ACTIVITIES THAT TEACH ATMOSPHERIC SCIENCE

OVERVIEW:

This unit gives students an opportunity to review meteorological concepts and to make simple weather equipment.

Pupils will see how the sun, wind, air pressure, humidity, temperature plus other factors fit into the hydrologic cycle.

Pupils build weather stations and construct weather maps. They experience the meaning and use of weather symbols.

WAYS TO MEASURE ENERGY ABSORPTION

The sun sends radiant energy to the earth in various amounts. This energy "triggers" atmospheric changes in the troposphere, the great mass of air near the earth. Scientists estimate approximately 43 percent of the total heat energy reaches the earth's surface.

Ask pupils what happens to the remaining 57 percent. Bring up the problem of atmosphere absorption, reflection, and scattering of heat energy. Discussion should include such factors as the earth's distance from the sun, the tilting of the earth's axis, absorption and reflection rates on land and water surfaces, atmospheric conditions, and the amount of solar energy released at one time.

ACTIVITY 37

Have pupils perform the following experiments:

Part A.

Fill a container three-fourths full of water. Take the temperature of the water. Set a light source (60-150 watt) approximately 8 inches above the container. Take thermometer readings every 2 minutes for 12 minutes.

Fill a second beaker (same size) with an equal amount of water. Take the temperature of the water. Set the light source approximately 12 inches above the container. Take temperature readings every 2 minutes for 12 minutes. Have students record their readings on a chart.

Part B.

Repeat Procedure A. Take readings every 2 minutes for 12 minutes with the light source directly above the container. Then repeat the readings with the light source moved several inches to one side of the container. Do not change the height of the light source.

Part C.

Gather 2 equal size containers. Fill one with soil, the other with water. Keep the levels of each container the same. Place both containers side-by-side directly under the light source. Take readings from each container every 2 minutes for 12 minutes.

Part D.

Take 2 equal size containers, blacken the outside of one container with candle soot. Leave the remaining container clear. Fill both with the same amount of water and place side-by-side directly beneath the light source. Take readings from each container every 2 minutes for 12 minutes.

Suggested Problems And Questions

1. Have students graph the results of experiments A through D. Tell them to place *temperature readings* on the *vertical axis* and *time* on the *horizontal axis.*
2. Assign pupils the following questions:
   a. How does the distance between the sun and earth affect how much heat the earth receives? The closer the sun, the warmer the rays providing these rays arrive in direct line from sun.
   b. How would the tilting of the earth's axis affect how much heat the earth receives? More tilt, less direct rays.
   c. Which surface seems to reflect more heat—soil or water? Why? Water. Provides more shiny surface.
   d. Which surface seems to absorb more heat—dark or light? Why? Dark. Offers least amount of reflection.
3. Ask students to extend the temperature readings for Activity 37 to see if any major changes occur.
4. Have pupils darken one side of a Florence flask and fill it with water. Tell them to turn the flask with the dark side facing away

from the light. Take 2 minute readings every 12 minutes. Repeat the above experiment. This time use a second flask with fresh water and turn the dark end side toward the light. Explain the results.

5. Have students stretch filter floss or a thin layer of cotton across a ringstand between a beaker filled with water and a light source. Take 2 minute readings every 12 minutes. Remove the floss and repeat the experiment.

Ask pupils to show how this activity simulates atmospheric conditions.

HOW TO TEST RATE OF EVAPORATION

Mention weather changes depend on water vapor in the air. How does this moisture enter the air? Water leaves the earth, penetrates the air, molecule by molecule, through the process of evaporation.

Tell pupils plants release large amounts of water vapor into the air. This process is known as transpiration. Water vapor helps insulate the atmosphere and prevents the loss of heat from the earth. This provides adequate moisture for plants and animals. Eventually the water vapor returns to the earth in the form of snow, rain, hail, etc. Use visual aids to help illustrate the hydrologic cycle. Stress water rises (evaporation) and develops into clouds or fog by cooling (condensation), then returns to earth as precipitation (solid or liquid form).

ACTIVITY 38

Emphasize high temperatures affect the evaporation rate of water; the higher the temperature, the more water will evaporate. Winds also cause considerable evaporation.

Let students discover which factors have the greatest effect on evaporation. Provide four glass Petri dishes for each group of students. Tell them to add 1 drop of water to each dish and test the following conditions:

1. Heat dish #1 over an alcohol burner or candle flame. Record the evaporation time.
2. Place dish #2 near a small electric fan. Record the evaporation time.
3. Set dish #3 under a light source. Record the evaporation time.

4. Place dish #4 on laboratory table. Allow drop to evaporate at room temperature. Record the evaporation time.

Have pupils construct a bar graph from their results. Summarize these findings with the class.

Suggested Problems And Questions

1. Ask students to test what effect, if any, the following conditions have on evaporation:
   a. Surface exposure of liquid (closed and open containers)
   b. Temperature of liquid
2. Have pupils test which liquid evaporates faster—tapwater, distilled water, or salt water (3 percent solution).
3. Have students test whether a liquid evaporates faster inside the room or outside the class. Ask them to list the conditions affecting evaporation rate.

WAYS OF MEASURING HUMIDITY

Stress humidity, the amount of moisture in the air, can be measured by a hygrometer. One variety, the psychrometer, operates on the principle of evaporation. Simply, the faster the evaporation, the cooler the temperature, and lower the reading on the thermometer. If evaporation is slow, the cooling effect is less.

Explain that two identical thermometers, wet and dry-bulb, measure relative humidity in the following manner: A cloth, cotton, or shoelace wick tied to a thermometer (wet-bulb) extends into a small dish of water. Water rising from the dish keeps the wick moist. Cooling of the wet-bulb thermometer (fanning with paper or card) causes evaporation from the wick. The relationship between wet and dry temperatures indicates relative humidity.

Mention relative humidity is an expression, written as a percent, of the amount of water actually present in the air compared to the maximum amount of water vapor the air can hold at that temperature.

ACTIVITY 39

Have students make a psychrometer. Provide thermometers, wicks (cotton, cloth, or shoelace), string, and ring stands. Offer these guidelines:

1. Attach a single-layer wick around the bulb of one thermometer. Dip the wick in a small dish of water.

2. Hang both thermometers from a ring stand. Keep them 2 or 3 inches apart. (Figure 19.)
3. Fan both thermometers with a stiff piece of paper or cardboard. Stop fanning the thermometers when wet-bulb thermometer reading remains steady.
4. Record the readings from both thermometers.
5. Find the relative humidity. Subtract the temperature readings on the wet and dry-bulb thermometer.
6. Check your readings against a relative humidity chart. Find the degree of difference between the 2 thermometers. Then locate the dry-bulb thermometer reading on the left side of the table, and the number directly in line with both of these readings. This will give the percent of relative humidity.

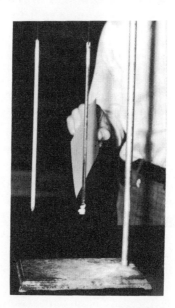

Figure 19

For example, if the temperature difference is 14°, and the dry-bulb reads 72°, the relative humidity is 42 percent.

Ask students to test the relative humidity for each of these areas:

1. Classroom floor
2. Laboratory table
3. Near the ceiling
4. Inside hall or passageway
5. Outside the classroom

ACTIVITY 40

Relate how human hair absorbs moisture from the air and increases in length. A clean hair (washed in alcohol) can be set up to stretch when the air becomes moist.

Allow students to explore ways for testing humidity with a human hair.* Provide them with small plastic beads, tape, pins, cardboard, scissors, beakers, burners, tongs, and other necessary equipment.

Here are some suggested problems:

1. By what percent does a moistened hair expand?
2. Test different animal hairs to see if they expand when moistened.
3. Test hair hygrometer at different levels—floor, lab table, and near the ceiling.

Demonstrations And Experiments

1. Dew point is the temperature when saturation occurs or when a given amount of vapor in a body of air will condense to form droplets.

    Let students determine dew point by filling a clean metal can three-fourths full of water. Wipe off the outside of the container, add ice, and stir with a thermometer. Dew point is reached when droplets begin to form on the outside of the can. Have pupils test the dew point inside and outside of the classroom.

2. Encourage students to make frost by filling a large metal container with crushed ice and rock salt. Tell pupils to alternate adding ice with salt, using twice as much ice. Frost will form on the outside of the container.

3. Pupils may also mix salt, ice, and water to form frost. Have students try different proportions of each ingredient to see which combination produces the lowest temperature.

4. Let students test which of the following combinations shows the highest evaporation rate:
    a. Soil and water
    b. Paper soaked in water
    c. Sponge saturated with water
    d. Sand in water
    e. Salt water

---

* Let pupils use their own hair. This helps increase student interest.

Tell them to weigh each container and ingredients before setting under a light source for 30 minutes. Keep all containers an equal distance from the light source.

5. Demonstrate dew point by adding dry ice instead of regular ice. Dry ice speeds up the cooling process. *Caution:* Do not handle dry ice with bare hands.

MAKING BAROMETERS FROM MERCURY, STRAW, AND MINERAL OIL

Mention air pressure, the weight of the atmosphere, changes with altitude; the higher the elevation, the less amount of pressure. Human ears are sensitive to this pressure change.

Relate that Torricelli, Italian scientist, invented the barometer, an instrument for measuring atmospheric pressure. Tell how Torricelli found mercury would rise approximately 30 inches in a tube free of air. He concluded air pressure pushing down on the dish of mercury caused it to rise in the tube.

Discuss the conditions which cause a barometer to rise or fall. Tell students a rising barometer generally indicates good weather; a falling barometer suggests inclement weather. Emphasize barometers measure changes in pressure force and are not intended to tell exactly the nature of coming weather.

ACTIVITY 41

Have pupils make a Torricelli-type barometer. Give them these guidelines:

1. Obtain several glass tubes approximately 36 inches in length.
2. Seal one end of the tube by heating over a Bunsen burner.
3. Fill tube with mercury.
4. Fill a small dish or beaker with mercury.
5. Close the open end of the tube with your thumb, invert, and place the lower end below the surface of the mercury. Remove thumb. The mercury will settle after a few seconds.
6. Make a cardboard scale. Attach on or near the glass tube. Mark the leveling point of mercury on the card.
7. Make daily readings. Record the mercury level changes in millimeters. Graph the results.

ACTIVITY 42

Allow students to make a straw barometer and instant barometer. Give them these directions:

A. Straw Barometer

1. Obtain a quart jar or large mouth container.
2. Stretch and fasten a thin layer of rubber (balloon) over the mouth of the jar.
3. Glue one end of a strand of straw to the center of the rubber covering.
4. Make a cardboard scale to indicate straw movement. High pressure pushing down on the rubber covering will cause the straw to point upward. Low pressure outside allows the inside flask pressure to push upward forcing the straw to point downward.
5. Make daily readings. Graph the results.

B. Instant Barometer

Provide students with mercury (or mineral oil), large test tubes or Erlenmeyer flasks, two-hole rubber stoppers, rubber siphons connected to glass tubing pieces, long barometer tubes (36 inches in length), stand and clamps.

Tell them to construct a barometer which will indicate pressure change when air is blown in or sucked out of the liquid-filled container. Give pupils enough information to get them started, but let them build their own instant barometer.

BUILDING COLORED WATER THERMOMETERS

Mention a thermometer measures the average speed of all moving molecules at a particular moment. This measurement indicates a change in temperature when a body adds or loses heat energy. Fast moving molecules increase the temperature reading and slow moving molecules lower it. The expansion and contraction of molecules greatly influences weather conditions.

ACTIVITY 43

Let pupils test the conditions which affect contraction and expansion. Have them do the following:

1. Fill a flask (Erlenmeyer or Florence) with colored water.
2. Use a Bunsen burner to heat a long glass tube. Bend into a U-shape. Note: heat tube for a short time only.
3. Insert the U-shape tube into a one-hole stopper, press the stopper into the flask, invert, and support on clamp and stand (Figure 20).
4. Observe the water level in the tube. Now place your hand over the flask. Record the results.

5. When the water level returns, cover the flask with moist, cold paper towels. Record the results.

ACTIVITY 44

Have students place a flask with colored water inside an empty beaker. Tell them to keep the neck of the flask above the water line. Add a long glass tube inserted into a one-hole stopper to the flask (Figure 21).

Have pupils test what effect, if any, the following conditions have on the water level in the tube:

1. Add ice water to the beaker.
2. Slowly pour hot water into the beaker.

Figure 21

Figure 20

3. Place warm, colored water into the flask. Submerge into a beaker of ice water.
4. Submerge a flask of colored ice water into a beaker of hot water.

Suggested Problems And Questions

1. Tell pupils to carry their flask thermometers outside (Activity 44) and measure temperatures next to the building, away from the building, and in a shaded area. Have them report their findings.
2. Ask pupils to experiment making Torricelli-type barometers with different lengths of glass tubing. Have them report their results to the class.
3. Give students various sized beakers and different lengths of rubber tubing. Let them investigate air pressure by setting up siphon displays. Demonstrate these points to get the class started:
   a. Fill a beaker with water. Elevate it several inches above an empty beaker.
   b. Fill a rubber tubing (approximately 12 to 15 inches long) with water. Hold both ends shut with your fingers.
   c. Place one end of the tubing in the beaker filled with water. Let the other end rest in the empty beaker. Release both fingers. Water will pour freely into the empty container.
4. Give students various diameters of glass tubing. Have them use their Bunsen burners to bend glass tubing to desired shape (Figure 22). Ask pupils to determine which of the following siphon shapes drains water the fastest:
   a. Long tube, wide diameter
   b. Short tube, wide diameter
   c. Short tube, narrow diameter
   d. Long tube, narrow diameter

Assign these questions:

1. How and where does air pressure affect the flow of water? Air pressure pushing against beaker water forces it up the siphon.
2. Where does air pressure affect the siphon arm the most? The least? The siphon end extended in the beaker of water. The siphon end touching the empty beaker.
3. Where does gravity affect the siphon arm the most? The least? Gravity pulls the water down out the longer siphon arm. At the top of the siphon arc (where it bends).
4. Why must you fill the siphon arm full of water before it works? So air pressure and gravity can do their work.

Figure 22

Demonstrations And Projects

1. Demonstrate air pressure. Fill a glass with water. Hold an index card firmly against the mouth of the glass. Invert the glass, and remove your hand. Describe how air pressure keeps the card from falling off.
2. Demonstrate air expansion. Tie a balloon around the mouth of an Erlenmeyer flask or empty bottle. Place the flask in a beaker one-third full of water. Gradually heat the water. What happens? Air heats up inside the bottle, expands, takes up more room, and forces the balloon to stretch.
3. Have students research the following problems:
   a. Why does sudden weather or pressure changes affect some people's sinuses?
   b. How does the Eustachian tube (which extends from eardrum to the throat) make adjustments for changing pressure?
4. Let students test whether a blond, black, or red strand of hair is affected most by humidity.

BRINGING CLOUDS INTO THE CLASSROOM

Introduce clouds through overlay transparencies, filmstrips, or color photos. Ask pupils what clouds are and how they form. List their responses on the board. Emphasize these points in your discussion:

1. Clouds are an accumulation of visible particles of water or ice.
2. Water particles form when the temperature reaches below dew point (above the earth's surface). The air's moisture changes from invisible gas to visible water droplets. In short, moist air rises, cools, and condenses.
3. Moist air particles form around tiny pieces of matter, e.g., dust, smoke, etc. These minute substances are known as condensation nuclei.
4. Clouds pick up, carry, and release moisture.
5. The higher air rises, the less the pressure becomes. Air expands as the pressure on it decreases. As air expands, it cools. If the air cools below its dew point, a cloud forms; rain or snow may fall.
6. There are 3 basic cloud types: cirrus, cumulus, and stratus. They are classified according to their shape and altitude. Their height depends upon latitude and time of the year.

Let students investigate the following cloud-making exercises:

ACTIVITY 45

Have pupils set up the apparatus shown in Figure 20. Note the moist paper towel draped over the upper flask. Tell them to fill the bottom flask one-half full of water, bring it to a boil, and carefully observe what forms in the inverted flask. Caution: Do not overboil. Allow the steam to slowly collect in the top flask.

Include these questions in the experiment:

1. How does this activity show the cloud-forming process? Cooling water vapor condenses.
2. Describe where evaporation, condensation, and precipitation occur. Evaporation takes place in the bottom flask; evaporation and condensation occur in the top flask.
3. What is the purpose of the moist towel covering? To cool the rising water vapor.
4. What determines how much moisture will collect in the upper flask? How much water evaporates.
5. What determines how long it will take moisture to form in the upper flask? The rate and intensity at which the water is heated.

ACTIVITY 46

Provide students with beakers (500-2000 ml), water droppers, calcium hydroxide, plaster of Paris, and methyl orange. Encourage them to make clouds by mixing the plaster of Paris or calcium hydroxide with water. Advise students to experiment mixing until their mixtures, when dropped in a beaker of clear water, drift slowly to the bottom and scatter particles in every direction. The pattern resembles a cumulus cloud formation. If students add two or three drops of methyl orange to their mixtures, a yellow-smog-like cumulus cloud appears.

ACTIVITY 47

Have students swirl hot water inside an Erlenmeyer flask (250-500 ml), dump it out, and fill the flask one-third full of hot water. Then tell them to set a piece of ice on top of the flask. A cloud forms inside the bottle when the hot water evaporates, rises, and cools upon reaching the ice. *Hint:* If the cloud is hard to see, turn off the lights, and shine a light against the flask.

CREATING WINDS IN A SHOEBOX

Begin this section by demonstrating air currents in a convection box (Figure 23). Replace the front of a cigar box or shoe box with glass. Cut two holes in the top of the box and place a glass chimney over each hole. Below one chimney place a lighted candle. Light a wooden splint or moist wad of paper and insert into the opposite chimney. The smoke will travel down the chimney, move across the box, and climb up the heated chimney. Have pupils observe what happens when the source of smoke is held above the chimney over the candle or the candle is moved between the chimneys or placed under each chimney.

Mention how this demonstration helps explain land and sea breeze formation. Bring up these points:

1. Wind is air in motion.
2. Unequal heating of the air causes winds.
3. The sun provides the energy to move the air.
4. The earth's rotation and the sun's heat create certain wind belts, e.g., trade winds, easterlies, westerlies and doldrums. Use a globe or world map to point out these patterns.

Suggested Problems And Questions

Assign pupils the following questions:

1. How does the speed of rotation effect circulation patterns? Drop dye into a spinning tray of water. Adjust the speed from very low

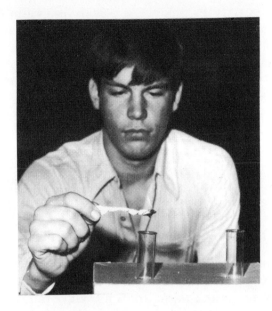

Figure 23

to very fast. Try different colored dyes for variety. Combine 2 or 3 different colored dyes simultaneously.

2. The Coriolis effect is a force created by the earth's rotation which deflects winds to the right in the northern hemisphere and to the left in the southern hemisphere. You can observe this effect on a world globe. Sprinkle powder or chalk dust over the globe. Slowly turn the globe counterclockwise and pour some water over the top. How does this activity demonstrate the Coriolis effect? Try colored water.

3. A wind vane indicates the direction of the wind. The arrow points to the direction from which the wind is coming. An anemometer measures wind velocity. Look at pictures showing these instruments.

   Make a combined wind vane—anemometer instrument. Use filter paper (fold into cups), wooden dowels, small nails, cement, cardboard, index cards, washers, and scissors. Make sure the wind vane balances properly and the anemometer turns freely.

4. What determines the character of an air mass? What decides whether an air mass remains over land or water, becomes warm and moist, warm and dry, cold and moist, or cold and dry?

5. How do moving air masses produce weather? Explain how the air mass with the greatest density, the coldest temperature, and the most weight dominates a particular area.

6. The surface between two air masses is known as a front. Tell what

conditions are necessary to produce a cold front, warm front, occluded front, and stationary front. Make diagrams or models showing these phenomena.

7. Look up cloud names, symbols, and descriptions. Try to make up stories which would support scientific findings. For example:

> When nimbostratus fills the air,
> plan to end your game;
> The weather is no longer fair,
> welcome snow or rain.

8. Make a model showing the three different levels of the troposphere in which clouds occur. Tell how the clouds in these levels differ from one another.

CONSTRUCTING MODEL WEATHER STATIONS

Discuss how air masses develop. Use Styrofoam or wooden blocks to show air movements. Tell students cold fronts usually bring brief storms and cooler weather; warm fronts introduce rain or snow; stationary fronts may bring long periods of precipitation; and occluded fronts bring precipitation.

Explain high pressure areas are large swirling air masses circling about a center of high pressure. Winds blow in a clockwise direction. They tend to bring clear weather and few clouds. Low pressure areas are large swirling air masses circling about a center of low pressure. Winds blow in a counter clockwise direction. They may bring rain or indicate a pressure system is approaching.

Introduce students to weather map symbols. Place an overlay transparency of a large area, state, or country on an overhead projector. Use a grease pencil to show how the following conditions are recorded by the weather bureau. (Pass out several blank maps of the area to be illustrated. Students can keep a record of these symbols.)

Conditions

1. Isotherms—Lines of equal temperature.
2. Isobars—Lines of equal pressure.
3. Cloud conditions
4. Cold front
5. Warm front
6. Stationary front
7. Wind speed and direction (Beaufort Scale)
8. Temperature

9. Rain, fog, frost, dew, drizzle, sleet, hail
10. High pressure
11. Low pressure

ACTIVITY 48

Part A

Pass out index cards and have pupils construct a station model (Figure 24) with the following information:

a. Sky, three-fourths covered.
b. Dew point, 37° F.
c. Wind—northwest, 15-20 mph.
d. Barometric pressure, 968.7 millibars.
e. Temperature, 66° F.
f. Visibility in miles and fractions, 1 1/4.
g. Amount of precipitation in last 6 hours, .35 (inches).
h. Symbol showing present state of weather (R).

Give them ditto sheets or text references which show standard weather bureau symbols.

Figure 24

Part B

Pass out blank area maps. Ask students to construct a simplified weather map which includes two different fronts, isobars, high and low areas, wind direction and speed, temperatures, and precipitation.

Part C

Pass out blank area maps. Below is a list of weather conditions and materials which students can attach (cement) to their maps. Have pupils create their own weather symbols and conditions.

Weather Conditions	*Materials*
1. High and Low Pressure (isobars)	1. String, thread, thin wire, colored pencils, pens and crayons.
2. Fronts (cold, warm, stationary, occluded)	2. Scissors and colored paper.
3. Cloud conditions	3. Circular hole punch, paper, colored pencils, straight pins, paper labels and straight pins with colored heads.
4. Temperatures (isotherms)	4. String, thread, thin wire, colored pencils, pens and crayons.
5. Wind conditions	5. Toothpicks, colored pencils or pens, pencils, and crayons.

Remind students to write a brief weather report to accompany their maps.

ACTIVITY 49

Let pupils work in groups and construct a meteorological station. Their stations should include the following:

1. A barometer
2. A thermometer
3. Hair hygrometer
4. Wet-and-dry bulb thermometer
5. Anemometer and wind vane

Ask students to make daily recordings of their instruments. Weather predictions and measurements from these crude instruments are not intended to be accurate. However, they do reveal the complexity of our atmospheric system and, hopefully, show students that the most experienced weatherman can miss on occasion.

EIGHTEEN DEMONSTRATIONS, PROJECTS, AND SUGGESTIONS FOR FURTHER STUDY

1. Tornadoes are violent storms over a narrow path with a whirling air mass forming a funnel-shaped cloud. Demonstrate a tornado by setting a large beaker of water over a magnetic stirrer. Add dye, a few drops at a time, to emphasize the whirling, funnel-shaped cloud.
2. Collect local weather forecasts for several days. Compare them with the actual weather.
3. Demonstrate rain. Fill a large beaker with ice and hold it over a flask of boiling water. The hot steam condenses into large drops of water which fall simulating rain.
4. Demonstrate the counter-clockwise motion of hurricanes. Fill a large beaker with water and swirl (stirring rod) the water in a counter-clockwise direction. Add 2 or 3 drops of food coloring.
5. A radiosonde is a radio-weather instrument attached to a large balloon which measures upper air temperature, pressure, and humidity.

   Students can have fun with the basic idea of upper air measurement. Have them make a parachute with a small block of wood attached (represents a radiosonde). Carefully wrap the parachute around the block of wood and attach to the back of a kite (Figure 25). This attachment must be made strong enough to become airborne yet loose enough to drop off the kite with a quick jerk of the string.

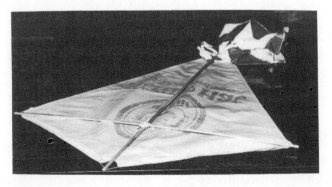

Figure 25

Send the kite high in the air. Give the string enough jerks to release the parachute. Recover the chute. Notice the distance the chute traveled with respect to kite height, wind speed and direction.

6. Sometimes it is necessary to convert temperatures from one scale to another. If you want to change fahrenheit to centigrade, use the formula:

$$C = 5/9 \, (F\text{-}32)$$

And to convert centigrade to fahrenheit:

$$F = 9/5 \, C + 32$$

Try the following problems: Convert to fahrenheit:
(a) $14°$ (b) $36°$ (c) $83°$ (d) $58°$
Convert to centigrade: (a) $87°$ (b) $11°$ (c) $39°$ (d) $-30°$ (e) $-6°$.

7. Write a report (with diagrams) showing how hail forms.

8. Report on cyclones and anticyclones.

9. Demonstrate lightning. Blow up two sausage-shaped balloons. Darken the room. Rub the balloons against your clothes (pants or dress). Bring them close together. Small sparkles of electricity race between the balloons.

10. Report to the class how thunderclouds develop. Tell the relationship between lightning and thunder.

11. Ice crystals take characteristic forms at different temperatures. The branched fern-like crystals which occur at approximately $-16°C$ can be simulated by preparing an ammonium chloride solution and putting a drop on a microscopic slide. Beautiful branching crystals appear.

12. Demonstrate a cold front. Place a glass partition in the middle of a glass tray. Fill one-half of the tray with salt water and the other half with fresh water. Add different colored dyes to the two halves. Remove the partition. The heavier salt water (cold air mass) moves under the lighter fresh water (warm air mass).

13. Demonstrate cloud seeding. Obtain a clean flask. Prepare a hot saturated solution of sodium thiosulfate. Pour the solution into another beaker. Let the solution cool. Add a sodium thiosulfate seed crystal to the solution. The sodium thiosulfate comes out of solution simulating rain from a cloud.

For additional study, give students the following topics:

1. Write a report on the United States Weather Bureau. Include the special services of the bureau.

2. Why do weathermen rely on a maximum and minimum thermometer? How do these thermometers differ?

3. Make a sling psychrometer. Mount two mercury thermometers so they can be swirled around. Wrap a thin piece of cloth in water. Leave the bulb on the other thermometer dry. Whirl both thermometers around rapidly. The cloth evaporates, cooling the bulb of the wet thermometer. Take the readings of both thermometers. The dry-bulb thermometer gives the temperature of the air. Take the difference in readings, and compare it to specially prepared relative humidity table.

4. List the things a pilot needs to know about the weather. What is meant by ceiling?

5. Report on attemps to control the weather. Include rain-making methods and ways to prevent storms.

chapter four

17 ACTIVITIES THAT
TEACH EARTH SCIENCE

OVERVIEW:

Students inspect minerals which make up the rocks that clutter the earth's surface. Pupils investigate individual mineral characteristics, e.g., specific gravity, hardness, streak, etc. They grow their own crystals and examine crystal growth under a microscope.

Special mineral features are considered. Activity 64 introduces a challenging game using the principle of double refraction. Activity 60, "Sherrock Holmes" offers a reward for uncovering the mystery mineral.

The unit closes with an evaluation of the three main rock groups: Igneous, metamorphic and sedimentary.

Begin by holding up a rock and asking the class, "Describe what I'm holding in my hand." The description should include shape, texture, color, weight, or outstanding feature. Write the student responses on the board. Pass around several different rock samples. Ask pupils how these samples differ from the first one shown. Tell them these samples help make up the lithosphere, the solid portion of the earth.

Mention how rocks are composed of one or more minerals and minerals are natural occurring inorganic (non-living) substances with definite chemical and physical properties. Reinforce this point by passing around mineral samples, e.g., pyrite, limonite, quartz, calcite, gypsum, mica, feldspar, etc. Emphasize that each specimen has its own unique characteristics.

ACTIVITY 50

Calculating Mineral Content

Relate how elements, substances that cannot be broken into simpler substances by ordinary chemical means, make up minerals which are the building blocks of rocks. Tell students ninety-two elements occur naturally on or near the earth. A few common elements make up the earth's crust, including the hydrosphere and atmosphere.

Give students the following information:

Element	Symbol	Approximate Percentage by Weight
Oxygen	O	49.7
Silicon	Si	25.9
Aluminum	AL	7.4
Iron	FE	4.4
Calcium	CA	3.3
Sodium	NA	2.4
Potassium	K	2.3
Magnesium	Mg	2.0
Others	—	2.6
		100.0

Prepare several lith-o-boxes. Place 50 g of colored paper punch (or plastic beads) into each container. Here is the recipe of paper or beads (by weight):

Element	Color	Weight (approximate)
Oxygen	Red	25 g
Silicon	Yellow	23 g
Aluminum	Blue	4 g
Iron	Black	2 g
Calcium	Green	2 g
Sodium	White	1 g
Potassium	Orange	1 g
Magnesium	Purple	1 g
Others	Lavender	1 g
		60 g

Have pupils match the color with the appropriate element. Let them use their information sheet for a guide. Make sure students know each container carries 60 g of paper. Note: You may have to shave paper pieces to equal some percentage weights.

WAYS TO GROW BEAUTIFUL CRYSTALS

Tell students crystals are naturally formed particles with regular shapes, flat surfaces, and straight edges. A mineral's internal structure which consists of atoms, ions and molecules, arranges itself in definite patterns. Thus, a crystal's form reveals a mineral's internal secrets.

Pass around minerals which represent various crystal shapes, e.g., halite, galena, pyrite, chalcopyrite, sulfur, quartz, calcite, feldspar, mica, olivine, tourmaline, etc. Briefly mention how crystal forms differ. A lengthy lecture describing axes intersections for each crystal system does little to stimulate interest.

ACTIVITY 51

What factors determine crystal growth? Let pupils experiment to find out for themselves. Give them just enough information to start. Begin with a saturated solution, a seed crystal, and a hint: Slowly evaporate the solution.

Hold a crystal-thon. Have students discover for themselves how to grow large crystals. Offer prizes, e.g., bonus points, etc., for outstanding results. Provide the following materials:

1. Beakers, jars, cans, Petri dishes, or cups.
2. Water.
3. Granulated sugar, table salt, copper sulfate, or Epsom salt.
4. String, wire, or thread.
5. Cheesecloth.
6. Small weights or washers.
7. Glass rods, wooden splints, or paper disks (for string attachment).
8. Stirrers.

Tell students to grow sugar, table salt, copper sulfate, or Epsom salt crystals. Remind them to label each container with student name, date, period and crystal solution. Allow crystals to grow from 7 to 10 days before deciding winners. Have pupils check their containers daily.

After announcing winners, discuss the things which regulate crystal growth. Some of these are:

1. The saturation level of the crystal solution.
2. Seed crystal size.
3. Temperature of crystal solution.

4. Size and shape of container.
5. Rate of evaporation. The slower the cooling, the larger the crystals.
6. Room temperature. Should be fairly constant.
7. The degree of crystal solution contamination.
8. Length and width of suspended string.
9. Size and depth to which weight was suspended in solution.
10. Length of time suspension remains undisturbed.
11. Uncontrollable factors, e.g., solubility of material, ability of substance to crystallize, etc.

ACTIVITY 52

Examining Crystals Under A Microscope

Give pupils the opportunity to prepare and examine micro-crystals. Provide them with microscopes, microscopic slides, lens paper, and eyedroppers.

Have them do the following:

A. Prepare an ammonium chloride solution. Place a drop of solution on a glass slide. Crystals resembling tree branches form.
B. Dissolve silver nitrate crystals in water. Place a piece of copper foil, wire, or copper penny on a glass slide. Place a drop of solution along the edge of the copper substance. Spike-like crystal patterns form.
C. Dissolve potassium dichromate in water until a bright orange color appears. Place a drop of solution on a glass slide. Yellow-orange twisting crystals form. Caution: Warn students silver nitrate and potassium dichromate are caustic and poisonous. Take special care in handling them.

These slides can be dried and stored without special treatment for future study.

Suggested Problems And Questions

Give interested students the following items:

1. A physical change is a change that does not produce a new substance. Does grinding a substance change its composition? Find out using salt or sugar.
2. A chemical change produces new substances. Show how heating an equal mixture of sulfur and iron illustrates this point. What are the individual characteristics of sulfur and iron? How does heating change this?

3. If commercial molecular kits are available, have pupils construct these compounds: water, hydrogen peroxide (H_2O_2), methane (CH_4), glucose ($C_6H_{12}O_6$), plus any other compound they choose to build. Students enjoy building molecules and may wish to continue the next day.

4. Calculate the percentage composition of a compound or determine formula from percentage composition. (See *Modern Road To Chemistry*, Bagby, Henry, Desjardins, pp. 119-125.) Students receive excellent practice in using the periodic table.

5. Choose several common earth elements. Make a chart showing each element's chemical symbol; electron, proton and neutron number; and atomic weight.

Demonstrations And Experiments

1. Obtain pictures or patterns of the 6 basic crystal systems (see H.H. Sisler et al; *General Chemistry*, Macmillan, 1949, and A. Holden and P. Singer, *Crystals and Crystal Growing*, Doubleday, 1960). Cut these patterns from paper or light cardboard, bend and tape the edges. Colored construction paper makes outstanding crystal patterns. Also, carve crystal shapes from balsa wood, potatoes, carrots, soap, or clay.

2. Collect various sized Sytrofoam balls, pipe cleaners, wire, toothpicks and cement. Construct crystal models of silicate minerals or silicates, i.e., silicon-oxygen tetrahedra, halite, and sodium chloride.

3. Set up a display of minerals and the crystal systems they represent.

4. A perfect crystal has faces which are alike in size, geometrical shape and appearance. Try to grow a perfect copper sulfate crystal.

5. Demonstrate thymol crystal growth. Slowly heat enough thymol crystals to cover the bottom of a glass Petri dish. Place the Petri dish on an overhead projector. Add 2 or 3 thymol crystals to the melt. Students will see crystals form from the melt. Caution: thymol mildly irritates the eyes and skin.

6. Demonstrate crystal models. Show the difference between lattice and packing models. Lattice structures show ion arrangements more clearly than packing models. They can be made with Styrofoam balls, wire, dowels, or toothpicks (Figure 26). Packing models show a tighter ion arrangement. They can be constructed by cementing marbles, Ping-Pong balls, or Styrofoam balls together (Figure 27). These models stimulate interest and promote discussion. They also reveal internal crystalline structures.

7. Place Epsom salt, table salt, copper sulfate and sugar crystals (produced in class) under a steroscopic microscope or hand lens. Sketch these crystals, and tell how each differs from one another.

HOW TO MAKE YOUR OWN ROCK

Some books suggest beginning a mineral study by handing out small pieces of granite to the students and having them identify the individual minerals. Granite pieces are generally too small to offer

Figure 26

Figure 27

individual mineral identification. Students may quickly lose interest. A more effective method would be to pass out large samples of the minerals which make up granite.

ACTIVITY 53

Pass out granite pieces to the class. Have pupils list the rock's physical characteristics: Texture, weight, shape and color. Next, distribute large samples of the minerals which compose granite, e.g., quartz, feldspar, hornblende and mica. Tell pupils to list all the physical characteristics they observe for each mineral.

ACTIVITY 54

Reproduce the work of Mother Nature in the classroom. Have students do the following:

1. Break up pieces of quartz, feldspar, mica and hornblende.
2. Mix several drops of a strong cement to the mineral pieces, stir, and pour ingredients into a Petri dish, jar lid, or small plastic container.
3. Let the mixture thoroughly dry.
4. If time permits, repeat the procedure with different portions of each mineral.

Assign the following questions:

1. Have you made granite rock? Explain. (No. Granite is an igneous rock which may be formed from metamorphic or sedimentary materials. Granite rock shows the effects of increased temperature and pressure. Activity 54 cannot duplicate this process.)
2. How does this artificial product compare with the original piece of granite? (In mineral composition only).
3. How does granite form? (Answered in question #1).
4. What is missing from your classroom product? (The process of molten material solidifying by cooling either above or below the surface of the earth. Also, the effects of temperature and pressure).

ACTIVITY 55

Let students make rocks which resemble granite, but differ in mineral content. Provide them with specimens of quartz, feldspar, mica, hornblende, pyroxene, biotite, augite, olivine and magnetite. Also, give them representative samples of granite (for comparison), syenite, diorite, diabase and gabbro.

The following chart tells students what minerals compose each rock. Let them use this information for a guideline. They need only repeat the procedure listed in Activity 54 to simulate rock samples. If rock samples are unavailable, provide pupils with text books illustrating these products.

Rock	Mineral Content
Granite	Quartz, feldspar, mica, and hornblende.
Syenite	Feldspar, mica, hornblende and little, if any, quartz.
Diorite	Hornblende or pyroxene, biotite-mica, feldspar and practically no quartz.
Diabase	Feldspar, augite, olivine, and magnetite.
Gabbro	Feldspar, pyroxene or hornblende, and no quartz.

WAYS TO TEST THE PHYSICAL PROPERTIES OF MINERALS

The unscientific statement, "All rocks look alike," is a poor evaluation and does not recognize the unique qualities of individual minerals which can be tested and observed in class.

Demonstrate how to test for mineral hardness, cleavage, fracture, streak, color (color is unreliable since some minerals display different colors), luster and specific gravity.

ACTIVITY 56

Streak reveals the color of the mineral when powdered, and is an important guide in identifying minerals. Give students streak plates from floor tile to scratch their minerals. Pass out mineral samples or rock collection kits. Have them compare external color with streak. For example, talc has a gray or green external color and a white streak.

ACTIVITY 57

Mineralogists refer to the surface appearance of a mineral in reflected light as luster. Pass out samples of shiny metal, imitation pearls or mother of pearl, satin, silk, glass, resin (amber) or rosin, wax, soil, and grease. Have pupils match their mineral samples with these items.

Hand out the following guideline information:

Luster or Shine	Product
Metallic	Silver, Copper, Aluminum, etc.
Adamantine	Oily surface (flashy, diamond-like or greasy)
Vitreous	Broken glass
Resinous	Rosin or Amber
Waxy	Pieces of Wax
Pearly	Imitation pearls or Mother of Pearl
Satiny	Silk or Satin
Dull or Earthy	Soil

ACTIVITY 58

Tell students specific gravity is a comparison of a mineral with water—a way of determining how many times as heavy as water a mineral is. For example, gold has a specific gravity of 19.3, nearly 20 times as heavy as water. (Refer to Chapter 6, *The Surrounding Sea,* Activity 92, for the method of determining specific gravity.)

Demonstrate how specific gravity is determined. Students may need help rounding off numbers or dividing decimals.

Have pupils test the specific gravity of several mineral samples. Select light, medium and heavy minerals. Here are some guidelines:

Mineral	*Specific Gravity (Approximate)*
1. Galena (Lead)	1. 7.5
2. Pyrite	2. 5.0
3. Talc	3. 2.8
4. Barite	4. 4.5
5. Quartz	5. 2.60 - 2.70
6. Serpentine	6. 2.5
7. Calcite	7. 2.7
8. Cuprite	8. 6.0
9. Dolomite	9. 2.8 - 2.9
10. Amphibole	10. 2.9 - 3.4
11. Feldspar	11. 2.5 - 2.7
12. Gypsum	12. 2.3

ACTIVITY 59

Relate that hardness is the resistance of a mineral to scratching. Show students the minerals which represent various degrees of hardness. These are:

Hardness	Mineral
1	Talc
2	Gypsum
3	Calcite
4	Fluorite
5	Apatite
6	Feldspar
7	Quartz
8	Topaz
9	Corundum
10	Diamond (if available)

Be sure students understand these numbers merely indicate the relative hardness. In other words, number 5 in the scale of hardness is not 5 times as hard as number one, etc.

Have students test the hardness of several mineral samples. Provide them with good steel knives or files and glass pieces. Students furnish their own fingernails and copper coins.

Here is the testing scale:

1,2— Minerals scratched by the fingernail have a hardness of 2.5 or less.

3— Minerals can be cut easily by a knife; just scratched by a copper coin; not scratched by a fingernail.

4— Minerals can be scratched by a knife without difficulty but not easily cut.

5— Minerals can be scratched by a knife with difficulty.

6— Minerals cannot be scratched by a knife; can be scratched by a file; will not scratch ordinary glass.

7— Minerals scratch glass easily.

8-10—Minerals seldom encountered for consideration.

Suggested Problems And Questions

Assign the following questions:

1. Tell the many ways minerals are used by man. (answers will vary.)
2. List 2 ways in which feldspar differs from quartz. (Feldspar cleaves with a flat surface that reflects light evenly. Quartz scratches glass easier than feldspar).

3. How does calcite differ from feldspar? (Hardness, all calcite sides are slanted, and calcite will effervesce when sprinkled with dilute hydrochloric acid.

4. Test several mineral samples by dipping a nichrome wire in an acid bath, transfer to a powdered mineral, and place over a Bunsen burner flame. How does this flame test help identify minerals? The elements making up a mineral give off a characteristic color, e.g., sodium—yellow; lithium—crimson, etc.

5. Find out which minerals effervesce in dilute, cold hydrochloric acid. Caution: Be sure to wash away acid after each test.

Demonstrations And Experiments

1. Demonstrate fluorescence and phosphorescence in minerals. Some minerals give off light rays when exposed to ultraviolet light. When they absorb the ultraviolet light and give off visible rays of longer wave lengths, they are said to be fluorescent. If a mineral continues to give off light after the ultraviolet rays have been cut off, they are said to be phosphorescent.

    Fluorescent minerals include willemite, calcite (some) and opal. Phosphorescent minerals are represented by willemite, calcite (some), sphalerite, uranium and tungsten.

2. Give students different size pieces of the same mineral. Ask them if size determines specific gravity, i.e., does a 20 g piece of quartz have the same specific gravity as a 60 g piece?

3. Have pupils grind up some gypsum, place in a test tube, and heat over a Bunsen burner flame. Ask them to examine the test tube and write down their observations. Tell them the formula for gypsum is $CaSO_4$, H_2O.

4. Pass out limonite samples. Have pupils grind them in a mortar, then examine the pieces. Ask them to report their findings. (Limonite is not a single mineral. It contains iron, oxygen and water.) Will heating drive the water out? Have pupils find out for themselves.

5. Let pupils grind up some pyrite, place in a test tube and heat over a Bunsen burner. Have them report their findings. (Few students will mistake the sulfur odor.)

6. Tell pupils to mix Kaolinite with water in a beaker. Ask them to test the odor, taste and feel. (Kaolinite has an earthy smell, sticks to the tongue, has a clay taste and greasy feel.)

PLAYING DETECTIVE: WHERE IS THE MYSTERY MINERAL?

ACTIVITY 60

Students enjoy playing the game "Sherrock Holmes." Give each group of students 8 different minerals. Tell them one of the minerals is worth 10 bonus points if correctly identified. Provide enough clues so pupils can uncover the mystery mineral. Give pupils free rein in testing each mineral.

Pass out these minerals: Calcite, quartz, hematite, feldspar, olivine, serpentine, pyrite and pyroxene. Tell students the bonus mineral has a specific gravity range from 2.5 to 2.6 (calcite, quartz, feldspar and serpentine); A hardness from 2.5 to 4.0 (calcite and serpentine); and a white streak (calcite and serpentine).

Problem: If these tests show either of two minerals (calcite or serpentine) could be the bonus mineral, what will be the determining factor? Give this final clue to only those pupils who realize either calcite or serpentine is the answer: The bonus mineral will effervesce readily in cold dilute hydrochloric acid. (Answer: Calcite.) Mix or match various minerals to keep students on their toes.

ACTIVITY 61

Break up three different minerals into small pebble size (4-5 mm diameter). Pour crushed material into jar lids or Petri dishes. Add a clear, strong cement, stir, and allow mixture to dry overnight. Pass out mineral clusters. Have students determine the specific gravity for the entire cluster and test the streak and hardness of individual minerals. Ask pupils to identify each mineral. Don't tell students how many minerals make up each cluster. Award bonus points for the correct answers.

DETERMINING SPECIAL PROPERTIES OF MINERALS

Mention how minerals possess special properties, e.g., earthy smell, color, crumbles easily, same color streak, fluoresce or phosphoresce under an ultraviolet light, etc.

ACTIVITY 62

Give pupils the following chart and mineral samples. Have them test each mineral as indicated on the chart and record their findings (possible answers included under Written Observation).

Mineral	Testing Procedure	Written Observation
Halite	Taste the specimen	Salty taste
Calcite	Place a clear crystal over a book, newspaper or magazine print.	See two images; known as double refraction.
Sulfur	Record the color and smell.	Yellow color. Sulfur smell.
Magnetite	Move a small magnet over the specimen.	Specimen is attracted to the magnet.
Quartz	Measure each crystal side with goniometer*.	All sides have the same angle.
Hematite	Rub against a streak plate.	Gives off characteristic brownish-red streak.
Kaolinite	Breathe on specimen. Test the odor. Touch tongue to specimen.	Has earthy odor. Sample tends to stick to the tongue.

* If goniometers are unavailable, students can connect two wooden splints at their base with a straight pin. This allows for easy movement and measuring of crystal sides (Figure 28).

Figure 28

ACTIVITY 63

Pass out samples of graphite, galena, cinnabar, chrysotile (asbestos), muscovite, olivine and talc. Let students study these minerals and list the special properties they discover for each one. Here are some observations:

1. Graphite	—	Soft; good for making marks or writing; has a metallic luster.
2. Galena	—	Has a bright metallic luster; cleaves easily; fairly soft; has a high specific gravity.
3. Cinnabar	—	Soft; very heavy; leaves a scarlet-red streak.
4. Chrysotile	—	Can be separated into very fine, flexible fibers; fibers have a silky luster.
5. Muscovite	—	Transparent to translucent; elastic and flexible; layers easily separated.
6. Olivine	—	Very hard; brittle.
7. Talc	—	Soapy feel; very soft.

SEEING DOUBLE WITH ICELAND SPAR

ACTIVITY 64

Relate how transparent calcite, "Iceland Spar," breaks up rays of light passing through it into two rays. Any object seen through it appears double. "Double Trouble," a calcite game, will create considerable interest.

Pass out large, clear calcite crystals (1 1/2" by 3/4"). Duplicate and pass out the following chart: (see page 102.)

Allow pupils to devise their own designs, problems,—and solutions to these problems. Tell the easily frustrated student to create a problem for someone else to solve (perhaps the teacher).

Suggested Problems And Questions

1. Ask students how they might tell the difference between chalcopyrite ("fool's gold") and gold. When chalcopyrite is pounded, it gives off a greenish-black streak and crumbles easily. Gold can be pounded, hammered, or pressed into different shapes without breaking. Also, gold has a specific gravity nearly 5 times as great as chalcopyrite.

Place crystal over this design	Turn crystal to make this pattern	Amount of time to solve problem
 ▲ ✕ ●	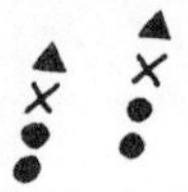 ▲ ✕ ● ▲ ✕ ●	
 ⅃	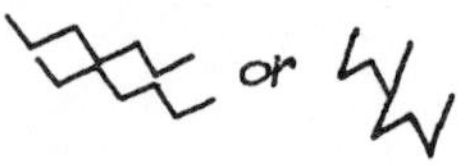 ⋉⋉ or ⅄	
 �8	 (jagged symbols)	
5	 (symbol)	

2. Certain crystals break in definite directions. Give pupils samples of galena, calcite, fluorite, feldspar and halite. Have them break these specimens into smaller pieces and sketch the fragments. Have students check whether or not all minerals break in the same manner.

3. Some minerals do not cleave along definite lines. Breakage may be uneven (irregular), hackly (jagged), or conchoidal (shell-like). Mineralogists call this fracture. Pass out samples of obsidian, flint and quartz. Let students break them apart and determine fracture.

4. Number the following mineral samples, pass them out, but don't tell students what they are: biotite, barite, pyroxene, sulfur, serpentine, feldspar and pyrite. *Problem:* Have pupils identify the minerals which show cleavage. (Answer: biotite, pyroxene and feldspar).

5. Spray with different colored paint several mineral specimens. Pass them out and have pupils identify them.

6. Hold a heavy weight contest. Let pupils find out which of the following numbered minerals have the highest specific gravity: 1)

galena (7.5), 2) cinnabar (8.1), 3) hematite (5.2), 4) barite (4.5), 5) magnetite (5.2), 6) cuprite (6.0) and 7) cassiterite (7.0).

7. Hold a middle weight contest. Give students these numbered minerals to test for specific gravity: 1) pyrite (5.0), 2) hematite (5.21), 3) barite (4.5), and 4) magnetite (5.2).

8. Hold a light weight contest. Pass out these numbered minerals and have pupils determine specific gravity: 1) calcite (2.7), 2) gypsum (2.3), 3) sulfur (2.0), 4) feldspar (2.5), 5) quartz (2.6), 6) serpentine (2.5) and 7) talc (2.8).

INSPECTING THE PROPERTIES OF ROCK

Spend time going over the three groups of rocks: igneous, sedimentary and metamorphic. (Ward's Filmstrip Series illustrates these rock-forming processes.)

Offer a brief description for each rock-forming process:

a. Igneous—Molten rock which cooled and solidified. Typical examples: granite, basalt, pumice and obsidian.

b. Sedimentary—An accumulation of rock particles which settle into horizontal layers and slowly unite together into rocks. These rock fragments may be deposited from wind, water, ice, or other means. Shale, sandstone, limestone and conglomerate are typical examples.

c. Metamorphic—Rocks whose original form have been changed by heat, pressure or chemical action. Typical examples: Slate, marble, quartzite and serpentine.

Rock classification is difficult because the rock-forming mineral combinations are many. Try this demonstration: Hold up a beaker of water. Tell students you will make granite—a combination of quartz, feldspar, with either mica or hornblende, or both. Add a drop of red dye for mica; blue for hornblende; green for quartz; and yellow for feldspar. Slowly stir the ingredients. Now ask pupils if they could easily identify this rock.

ACTIVITY 65

Hand out the following chart: (see page 104.)

Punch out pieces of different colored paper (or use beads) to represent these minerals:

1. Quartz—white
2. Feldspar—red
3. Plagioclase feldspar—orange

Igneous Rock	Composition
Diorite	Little, if any quartz. Plagioclase feldspar, hornblende, biotite and pyroxene.
Peridotite	No feldspar. Contains hornblende, biotite and pyroxene. Olivine conspicuous.
Gabbro	Chiefly plagioclase and pyroxene. Biotite and hornblende present. Olivine conspicuous.
Diabase	Feldspar, augite, olivine and magnetite.

4. Hornblende—green
5. Biotite—black
6. Pyroxene—blue
7. Olivine—yellow
8. Augite—pink
9. Magnetite—brown

Fill and number several small jars or test tubes with 30 grams of the following:

Jar 1 (Diorite)
* White paper (quartz) — 2 g
* Red paper (feldspar) — 7 g
* Green paper (hornblende) — 7 g
* Black paper (biotite) — 7 g
* Blue paper (pyroxene) — 7 g
                          30 g

Jar 2 (Peridotite)
* Green paper (hornblende) — 6 g
* Black paper (biotite) — 6 g
* Blue paper (pyroxene) — 6 g
* Yellow paper (olivine) — 12 g
                          30 g

ACTIVITY 66

Provide students with these rock samples: Granite, felsite, basalt, talc-schist, slate, gneiss, marble, sandstone, limestone, shale, and quartzite. Ask pupils to separate the rocks into their respective groups: igneous, sedimentary and metamorphic.

Tell students that generally speaking, metamorphosed igneous rocks show banding of rearranged minerals; metamorphosed sedimentary rocks usually display crystalline structures. Remind pupils how these three rock groups are formed. Encourage them to test individual mineral properties, break the rocks apart and observe under a microscope or magnifying glass. Give pupils available rock collections or text books for references.

TWELVE SUGGESTIONS FOR FURTHER STUDY

Assign these problems to students wishing extra work:

1. Pressure within the earth's crust can cause the atoms that constitute the surrounding rock to change positions and form new crystalline structures.

   Tell pupils to knead glitter or crystal decors throughout a ball of plasticine clay. Then press the clay into a one-inch layer, slice it, and examine its texture. How does pressure affect mineral alignment? Try pressing the clay at different angles with various force.
2. Use the hardness test to determine the hardness of several igneous, sedimentary and metamorphic rock samples. Which specimens are the hardest? Why?
3. Tenacity, the way a mineral stays together, can be tested. If a mineral is *malleable*, it can be flattened out without breaking. If it is *ductile*, it can be drawn out into a wire. A *sectile* mineral can be carved like a block of wood. A *brittle* mineral crumbles easily; an *elastic* one will bend and remain that way.

   Tell students to test the tenacity of these examples: A lead fishing weight, copper penny, sulfur, limonite, muscovite, selinite, gypsum, talc, apatite, graphite, pyrite, serpentine and chrysotile.
4. Diagram the Bowen Reaction Series. What does this series indicate?
5. What are ferromagnesian minerals? How are they important to rock identification?
6. List eight common minerals of igneous rocks. Briefly describe each one.
7. Describe rock texture. Include these terms: phaneritic, aphanitic, glassy, porphyritic and fragmental. Give examples for each of these.
8. Which rocks have a higher specific gravity, dark-colored or light-colored rocks?
9. Which rock type is generally heavier: igneous, sedimentary or metamorphic?

10. Give pupils several capsules containing small pieces of igneous, sedimentary and metamorphic rock. Have them identify each mineral, measure size (mm.), describe the shape and texture, and report the relative abundance.
11. Tell what materials are necessary to identify different rocks and minerals in the field.
12. Have students prepare rock samples showing relationship of metamorphic to igneous and sedimentary rocks.

chapter five

20 ACTIVITIES THAT TEACH ABOUT THE EARTH— FROM "CORE TO CRUST"

OVERVIEW:

Students examine the earth's diastrophic moods. They encounter activities which stress expansion, contraction, warping, buckling and twisting.

Pupils make model cities (blocks and clay) and a seismograph to measure "earthquake" shock. Blocks resembling building structures tumble with each induced shock. Students survey the aftermath and calculate the damage.

The volcano section includes 5 activities which emphasize volcanic rocks, eruptions and laval flows.

Pupils explore how organisms become fossils. They either duplicate fossil formation from alcohol, molasses, sugar and syrup; or make "fossils" from wax, lead, and plaster of Paris.

The unit closes with students exploring map making and interpretation. Activity #86 offers a unique way to make a map from water and paint.

WAYS TO PRODUCE EARTH MOVEMENT

Begin by mentioning how the earth remains unsettled. Relate how volcanic eruptions and earthquakes help our planet find relief from the forces of stress and strain located within its crust.

Demonstrate with different layers of colored clay how rocks shear and compress. Show how forces push against a body from two directions and produce shearing. Illustrate how forces touch a body from directly opposite sides and compress it (Figure 29). These forces bend, twist and break the earth's crust. Tell pupils how crustal movement, or diastrophism, sculptures the land in many ways. Use transparencies, slides, filmstrips, blackboard diagrams, or models to show folding, i.e., anticlines and synclines, subsidence, and other mountain building processes.

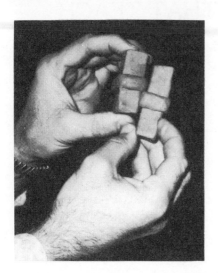 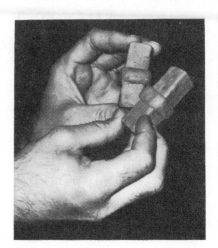

Figure 29

ACTIVITY 67

Hand out clay, wood blocks, glitter and silver tinsel. Ask students to make models of diastrophic events. Tell them to devise ways to use the glitter and tinsel for demonstrating crustal movement.

ACTIVITY 68

Ask pupils what forces cause diastrophism. Some may say unequal pressure and temperature, or the release of energy from deep within the earth's crust. Allow them to test their theories.

Part A

Pass out aluminum foil, clay and pestles. Problem: Show how an upward force changes a mountain structure. Suggestion: Cover a small mound of clay with foil. Mold a mountain. Measure its diameter and height. Slowly push the pestle through the clay bottom in an upward direction. Have students measure the diameter and height change.

Part B

Fill a small, round balloon with water. Seal the mouth tightly. Place in freezer overnight. Remove, place in tray, and observe what happens as the water thaws.

Part C

Attach a small balloon to the mouth of an Erlenmeyer flask containing 50 cc of water. Slowly heat the water. Remove the heat source before the balloon explodes. Place flask in a container of cold water. How does this experiment relate to diastrophism? Students should see heat causes expansion, cooling produces contraction, and if enough force is applied, the balloon explodes.

ACTIVITY 69

Recipe for a Pancake Earth

Spend time reviewing the contraction, convection, expansion, and continental drift theories. This two-part experiment will help students grasp these ideas:

A. Part I—The Effect of Heat on Cooler Subtances

Materials:

Pancake flour or Bisquick	Hot plate
Pyrex or metal tray	Magnifier
Teaspoon	Metric ruler
Water	Beaker (100 ml)
Thermometer (F°)	

Procedure: Mix 4 teaspoonsfuls of pancake flour with water in a 100 ml beaker. Add enough water to get a pasty substance. Stir until the flour mixes thoroughly with the water. Pour the mixture into a Pyrex dish or metal tray. Spread around in a circle. Place a thermometer (F°) in the mixture and record the temperature. Measure (mm) and record the diameter of the mixture. Place these recordings in the spaces provided on the Part I Observation Chart. Transfer the tray of doughy substance to a hot plate. Set the temperature dial at 350° F.

Part I
Observation Chart
The Effect of Heat on Cooler Substances

Structural Changes in Mixture
(Check ✓ appropriate spaces below)

Time (Min.)	Temp. (F)	Gas Bubbles Present	Surface Becoming Dry	Cracks Forming	Folds Develop- ing	Holes Forming	Rapid Rising & Falling of Dough	Rapid Harden- ing of Surface	Very Little Change
0									
2									
4									
6									
8									
10									

Diameter of mixture before
heating _____ mm

B. Part II—The Effect of Cooling on Warmer Subtances

Procedure: After 10 minutes, remove the mixture from the hot plate. Continue to observe dough temperature and texture changes every two minutes for 20 minutes. Make appropriate checks on the Part II Observation Chart. Allow mixture to stand overnight. The following day make final observations of structural and temperature changes.

C. Part III—After 24 Hours

1. Once more measure (mm) the length and width of the cracks previously measured in Part II. Have they changed in size? If so, how?
2. Does this experiment provide evidence supporting the contraction theory? Explain.
3. Has the mixture's diameter changed over the past 24 hours? If so, what does this indicate?

Have pupils answer the following questions:

Part I

1. How does heat affect the mixture?
2. What was the mixture's temperature after 10 minutes of heating? How much did it increase over the starting temperature?
3. When do bubbles appear in the mixture?
4. When does the surface begin to dry?
5. Describe how the following structures form in the mixture:
Plateaus	Mountains
Plains	Volcanoes
Ocean basins	Faults
6. When do each of the above structures appear (minutes)?
7. Does this experiment provide evidence supporting the convection, continental drift, or expansion theory? Explain.

Part II

1. How does cooling affect the mixture?
2. When did cracking occur in the mixture? Measure (mm) the length and width of the largest cracks. Sketch their approximate position.
3. When did the mixture's surface appear dry?

Part II
Observation Chart
The Effect of Cooling on Warmer Substances

Time (Min.)	Temp. (F°)	Structural Changes in Mixture (Check ✓ Appropriate Spaces Below)							
		Gas Bubbles Present	Surface Becoming Dry	Cracks Forming	Folds Developing	Holes Forming	Rapid Rising & Falling of Dough	Rapid Hardening of Surface	Very Little Change
12									
14									
16									
18									
20									
22									
24									
26									
28									
30									

Diameter of mixture after heating and cooling _____ mm

Overnight

Time (hrs.)	Temp. (F)	Holes Grew Larger	Shrinking Occurred	Cracks Increased In Width and Length
24				

Diameter of mixture after 24 hours _____ mm

4. How much did the temperature decrease after 20 minutes of cooling?
5. Does this experiment provide evidence supporting the contraction theory? Explain.

BUILDING A MODEL SEISMOGRAPH

Introduce the subject of earthquakes. Since most pupils seem fascinated with the devasting effects of earthquakes, tell them stories and show pictures of crustal destruction. (See Adams, *Earthquakes,* D.C. Heath and Company, Boston, 1964.) Stress terms pupils will encounter, e.g., Richter scale, seismogram, seismograph, seismic waves, tremor, fault, focus and epicenter. Briefly describe the following vibration patterns:

1. P waves or primary waves—Compressional waves which move longitudinally through liquids and solids. These waves set particles vibrating back and forth along the line of travel. Demonstrate by holding up a spring and snapping one end.
2. S waves or secondary waves—Transverse waves which travel through a solid medium only. The waves set particles vibrating at right angles to the line of travel. Demonstrate by moving one end of a rope up and down.
3. L waves or surface waves—Surface waves which travel along the earth's surface from the epicenter. These cause the damage produced by earthquakes.

ACTIVITY 70

Have students make a seismograph. Show them pictures or sketch several types of seismographs and briefly explain their function. Provide pupils with these materials: Scrap wood, springs, wire, weights, thread, rubber bands, empty metal or wood thread spools, candles or carbon powder (for coating thread spools), doweling, pins, string, cement and ringstands. Objective: Make a seismograph which will record vibration.

ACTIVITY 71

Tell pupils to make a small city. Include toy cars, animals, people, vegetation, etc. Houses and buildings can be made from clay or wooden blocks. Have students produce strong, medium and low intensity vibrations at different locations around their models. Ask them to record their results. Discuss these findings in class.

Demonstrations And Experiments

1. When enough energy accumulates rocks either break or snap back. This sudden release produces earthquakes. Demonstrate by holding a two-inch strip of clay over a transparency projector. Slowly pull the strip apart. Measure how far the clay stretched.

2. Demonstrate sudden expansion by placing ice cubes in a beaker of water. The sudden change in temperature causes the ice to crack.

3. Give students several cubes of ice. Ask them to examine the ice and estimate the location and length of fracture for each piece. Then place the cubes in water. How accurate were their predictions? Discuss these findings in class.
   Experiment questions:
   a. Will ice break in cold water?
   b. Will all ice cubes crack?
   c. Which cubes seem to demonstrate the greatest degree of cracking? (Large ice cubes, small ice cubes, etc.)
   d. Which ice cubes seem to crack the most—dirty (made with sandy or dirty water) or clear?
   e. Does shape determine direction and degree of cracking?
   f. How does this experiment relate to earthquakes?

4. Geologists say the earth is composed of different materials in the crust, mantle and core. Earthquake wave travel time and direction is determined by this factor. Hand out small Styrofoam balls and clay. Have pupils mold an outside clay cover around the ball. Tell them to slowly insert a thin nail through their models. Ask them how this activity relates to P wave movement. Students should see P waves experience less resistance in less dense material and therefore travel faster.

TESTING THE POWER OF SHOCK WAVES

ACTIVITY 72

Describe how the Galitzin-Kirkpatrick scale measures earthquake intensity. Mention how round, wooden blocks of the same diameter, but different lengths, are placed on a flat, horizontal surface. Soft vibrations cause the taller blocks to fall first; more intense shocks are required to knock down the smaller blocks.

Part A.

Pass out round, wooden blocks with the same diameter, but different lengths. Tell pupils to randomly place the blocks over a flat surface (desk or lab table). Then hit the surface, easy at first, and record which blocks fall with each successive shock. For example, the following chart shows intensity and damage: (Let pupils make their own charts.)

Intensity	Damage (fallen blocks)
1	1 long, thin block
2	1 medium block 2 long, thin blocks
3	1 long, thin block 2 medium blocks
4	1 medium block 2 short blocks
5	2 short blocks (Total Destruction)

Part B.

Have students set up their blocks, exchange intensity charts with fellow classmates, and try to duplicate the results by matching shock waves with damage. Play match-a-shock. The first person who correctly matches the intensity with the damage wins.

INSPECTING VOLCANIC ROCK

Volcanoes have a way of activating the senses and bringing a magma chamber of fun to the surface. The following suggested activities help students understand the explosive side of Mother Nature.

ACTIVITY 73

Mention that volcanic rocks differ according to their mineral content. Whether or not they contain quartz, feldspar, olivine, hornblende, biotite, pyroxene, or other mineral products depends upon where the volcano decides to erupt. And its regurgitants may be coarse or fine-grained, porphyritic (a texture of fairly large crystals set in a mass of very fine crystals), or glassy depending upon cooling rate.

Have students inspect volcanic rocks which differ in shape, weight, texture, and mineral content. Pumice, obsidian, scoria, basalt, granite, diorite, and gabbro make excellent examples. A dissecting microscope or good quality hand lens brings out rock structures in finer detail. Ask students to sketch what they see.

HOW TO MAKE A "MAGMA CHAMBER" AND LAVA FLOW

ACTIVITY 74

Relate that no one can say exactly what forces encourage magma to belch its way to the surface. Scientists believe the earth's original heat, radioactive elements in the mantle, and/or large rock movements in the earth's lower crust generate enough energy to force molten magma to the surface.

Whatever their origin, the rising gases, ash, hot cinders, and air-borne fragments bring destruction to practically all forms of life. Gas bubbles trapped in the lava (molten materials flowing over the earth's surface) expand producing holes in the fragments.

Part A.

Have students place a paper match under the microscope, examine the rough texture of the head, and sketch what they see. Then tell them to strike the match, let it burn for approximately 5 seconds, blow out the flame, and place it under the microscope. Ask them to make another sketch and answer the following questions:

1. Do escaping gases change the texture of the match head? If so, in what way? Yes, creates additional spaces throughout the rock.
2. List some volcanic rocks which show gas bubble chambers. Refer to Activity 73. Scoria and pumice are good examples.

Part B.

This experiment simulates the behavior of magma, which rises to the earth's surface, gives off gas, and settles as lava encrusted with holes. Instruct pupils to do the following:

*   Put 80 g of table sugar in a 150 ml Pyrex beaker.
*   Pour 20 ml of concentrated sulfuric acid over the sugar (this should be done by the instructor). The yellow sugar-acid solution

begins to effervesce, turn black, and slowly rises. The cylindrical carbon core, sometimes referred to as a carbon "snake," crawls up the wall of the beaker, and reaches a length of several inches.

Activate the student's imagination by suggesting the carbon mass represents magma which changes to lava when it passes the mouth of the beaker.

* (Optional) Have students weigh the beaker, sugar, and acid before and after the reaction.

Assign these questions:

1. Do they weigh the same? No.
2. How do you account for any difference in weight? Some weight loss due to escaping gas.

Tell pupils to examine and sketch their carbon products. Save all materials for Part C.

Part C.

Mention diamonds have been found in what some scientists think are the necks of ancient volcanoes. These pure carbon gems are exceedingly hard and believed to be formed under conditions of extreme heat and pressure.

If the student examines the bottom of the beaker containing the carbon mass, he'll notice not all the sugar has reacted with the acid. Some crystals remain unchanged and nestle comfortably against carbon particles. Have pupils scrape off these crystals (with attached carbon particles) and place them under the microscope (Figure 30). The examined specimens resemble priceless gems (hopefully, "diamonds") ready to be discovered.

ACTIVITY 75

Describe the two common types of lava flows: Pahoehoe or ropy, and AA or blocky. A mixture of mortar and water oozing over rocks resembles the ropy type of flow.

Have students make their own pahoehoe formation. Tell them to:

* Fill a 250 ml beaker with dry mortar to two-thirds volume.
* Slowly add water until the mixture has a consistency of a thick milk shake.

Figure 30

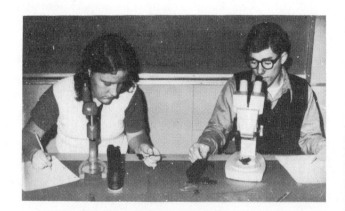

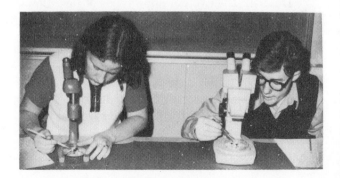

* Pour the mortar over a pile of small boulders or down a glass tray tipped at a slight angle.
* Allow mixture to stand overnight.

Answer these questions:

1. Where does the ropy pattern seem to be more pronounced, at the top, along the sides, or at the bottom of the flow? Why? At the bottom where mixture meets the greatest resistance.
2. Do blister-like swellings (cracks) develop along the lava crust? If so, where? How do you account for these cracks? Answers will vary.

MEASURING THE ANGLE OF REST

ACTIVITY 76

Mention that the angle of the steepest slope upon which loose materials can remain is called the angle of rest. This angle depends upon the type of material being piled up.

The form of a volcanic mountain depends upon several factors: The size, weight, and texture of rocks, speed of lava flow, and the nature of the volcanic eruption—i.e., whether or not it is violently destructive.

Ask pupils to do the following:

* Make a measuring device by fastening two wooden splint ends together with a straight pin.
* Slowly pour iron filings, charcoal powder, boiling chips, sand (fine or coarse), plaster of Paris, ground chalk (gypsum powder), sugar, sawdust, or soil on a piece of paper. Pour one material at a time.
* Measure the steepest angle by sighting along the desk top at eye level. (Figure 31.)
* Sketch each angle and estimate its size.

Figure 31

Assign these questions:

1. Which materials produce the steepest angles? Coarse materials.
2. What volcanic materials do you think form steep angles of rest? Those with rough textures.

MOLDING A VOLCANIC TERRAIN

ACTIVITY 77

Show the class pictures (filmstrips, overlay transparencies, slides, or films) of the main types of volcanic mountains: Shield-shaped domes, composite cones, cinder cones, lava domes, and spatter cones. Stories of destruction (Mount Vesuvius, Paricutin, etc.) help generate interest.

After discussion give each student a ball of clay, plastic Petri dish, toothpicks, nails, or wood splints. Have them build a volcanic landscape. Offer bonus points or prizes for the top three sculptures. Give students these guidelines:

1. Make as many different volcanic structures as you can.
2. Include ropy and blocky lava flows.
3. Add a realistic touch. Include soil, sand, iron filings, or any other material to your clay.

Demonstrations And Experiments

These classroom demonstrations are quite impressive. Tell students they are not designed to show exactly how volcanoes erupt. They simply reveal sparkling colors, bundles of smoke and hissing noises. Darken the room for full effect.

1. Put approximately 40 g of ammonium nitrate and 10 g of ammonium chloride in a mortar. Slowly grind to a fine powder. Add 10 g of zinc powder. Mix together. Cover the demonstration table with paper. Place mortar near the center of the table. Place a drop of water (eye dropper) in the center of the mixture. A loud, hissing sound shoves a smoke cloud toward the ceiling. A spray of colors fill the air. A light, yellowish brown ash remains behind. (Caution: Keep all students a safe distance from the demonstration table.)
2. Put approximately 30 g of powdered sulfur and 30 g of powdered zinc in a large mortar. Mix together. Insert a 3 inch magnesium ribbon strip into the center of the mixture. Avoid leaning the ribbon against the mortar. Cover the demonstration

table with paper. Light the magnesium ribbon. Stand back. A violent reaction occurs. Hot magnesium and sulfur particles fly through the air, drift back to the paper, and start miniature fires around the mortar.

3. Fill a small crucible with fine-grained, ammonium dichromate crystals. Insert a 3 inch magnesium ribbon strip into the center of the mixture. Avoid leaning the ribbon against the mortar. Cover the demonstration table with paper. Light the magnesium ribbon. Stand back. The reaction produces a gurgling sound, orange glow, wisps of smoke, and extrudes a light green ash which slowly surrounds the crucible. The finished product resembles a cinder cone.

    If students make their own volcanoes, have them carefully weigh the crucible, ammonium dichromate and magnesium ribbon. After the reaction, have them collect and weigh the ash. (Caution: Safety glasses should be worn.)

4. Have pupils make a "Soda Cone." Tell them to fill a 250 ml Florence flask one-third full with bicarbonate of soda. Add charcoal powder to turn the soda black. Cover the lab table or desk with paper. Add enough vinegar to cause the reaction to flow out of the flask, down the sides, and onto the paper. Carbon dioxide gas brings the soda and charcoal particles out of the flask. Ask students to trace (pencil or pen) the flow pattern on their paper.

    Will the reaction occur again without adding more vinegar? Ask students to restir the ingredients. A second eruption pours more "lava" onto the paper. Again, have pupils trace the flow pattern. How many times will the volcano erupt? Let pupils find out for themselves.

5. Let students make a clay volcano. Have them mold clay around a 100 ml beaker containing 70 g of table sugar. Shape a steep-sided volcano around the sides of the beaker. Leave a hole in the top large enough to insert the stem of a glass funnel. Pour 20 ml's of concentrated sulfuric acid into the funnel (teacher should pour the acid). The "lava" slowly creeps through the mouth (vent) and down the sides. Some material forces its way through the sides of the model (usually the weaker zones). Peel the clay away. Describe the magma chamber near the top of the volcano. (Optional) Repeat the above experiment using a 250 ml Florence flask. Place paper or tray under the flask. The reaction sputters and sends small gobs of carbon flying out of the flask. The remaining mass flows down the neck of the flask. Spatter cones put on a similar display.

CREATING "FOSSILS" FROM WAX, CLAY, LEAD, AND SYRUP

Open a fossil unit by showing how organisms become fossils. Film strips (Ward's Fossil Series), slides, transparencies, or actual examples help illustrate the following points:

* Fossils are remains or traces of organisms that lived during ancient geological times.
* They were buried in rocks that accumulated in the earth's crust.

Emphasize the term "ancient geological times" embraces all earth history from the earliest ages recorded in rocks to the epoch that immediately followed the last Ice Age—a range exceeding 3 billion years. Also, the word "rock" describes any extensive deposit that makes up part of the earth's outer portion, including ice, gravel beds, and clay.

The following activities help students recognize factors directly involved in the fossilization process.

ACTIVITY 78

Leaves, stems, or shells make excellent external molds in almost any kind of sedimentary rock. Shales, sandstones, and limestones house fine examples. Students enjoy making imprints using candle wax, food dyes or colored crayons, beakers, burners, glass Petri dishes, tongs, and plasticene clay. Provide them with these directions:

1. Spread clay over the Petri dish.
2. Press a twig, leaf, or shell into the clay.
3. Melt wax in a beaker. Add a colored crayon or dye to the wax (dark colors work best).
4. Remove beaker from the burner. Pour molten wax over specimen in clay. Be sure to use enough wax to cover the entire specimen.
5. When the wax cools, carefully remove the clay. Label your finished product (Figure 32). A thin film of Vaseline spread over the "fossil" helps keep it from sticking to the clay.

The shape and veins of the leaves show outstanding detail; the dye is suggestive of mineral matter. Shells pressed into clay pads provide outstanding molds.

Tell pupils the difference between a cast and a mold. A mold is the remains of an organism which, after decaying or dissolving, leaves a cavity. A cast is the filling of a mold with wax, plaster of Paris, lead,

Figure 32

or other substance which duplicates the shape and texture of the fossil.

   Carbonization, a process in which organisms become partly decomposed and leave a residue of carbon, can be demonstrated if students pour black wax over their leaves. Black wax is easily produced when carbon powder is added to melting wax.

ACTIVITY 79

   Describe permineralization. Explain (pass around available specimens) that cell materials in bone, teeth, or shell gradually disappear as mineral water soaks into the cell space. The water evaporates and leaves behind a layer of solid material lining each cell cavity. The process of soaking and drying is repeated until the original cells become completely replaced. Some of the petrifying minerals are calcium carbonate, pyrite, and silicon dioxide.

   Let students make "instant permineralized fossils." Provide

them with lead, plaster of Paris, glass Petri dishes, tongs, stirrers, beakers, metal cups or pots (for melting lead), Bunsen burners, and assorted shells. Give students these guidelines:

1. Mix plaster with water until a thick paste forms.
2. Pour pasty substance in several Petri dishes, insert shells and let dry overnight. Do not sink shells too deeply into the plaster.
3. The next day carefully remove the shells by gently prying around the edges.
4. Melt the lead and pour into the plaster molds. Note: Use only small amounts of lead.
5. When the lead cools it will contract enough to be easily removed from the plaster molds. The plaster molds and lead casts make striking displays (Figure 33).

Figure 33

ACTIVITY 80

Relate how amber, a fossil gum from the sap of ancient plants, trapped unsuspecting insects. Amber became an eternal coffin by capturing, sealing, and preserving insects. The amber would harden, drying the creature to almost nothing. But, in many cases, remains reveal heads, antennae, wings, bodies and legs.

Pupils can make their own "amber." Have them do the following:

1. Mix 2 parts sugar, 1 part Karo syrup, and 1 part water in a beaker.
2. Boil solution for approximately 10 minutes. Let pupils experiment until the solution reaches the right "cracking point," i.e., becomes

brittle when dipped in cool water. Add molasses, a drop at a time, until the desired amber color is reached.

3. Remove the ingredients from the burner. Pour into a Petri dish or shallow tray.

4. Drop several dead insects into the substance. Arrange the insects in different positions throughout the liquid. When the substance cools it hardens and encases the insects. Trapped air bubbles add realism to the final product. (Figure 34.)

Figure 34

ACTIVITY 81

Describe how the Rancho La Brea tar beds, Hancock Park, Los Angeles, California trapped Pleistocene animals—sloths, dire wolves, saber-tooth cats, mountain lions, coyotes, bisons, birds, etc. Scientists believe these animals drank from the water floating above the liquid tar, became stuck in the asphalt, and slowly sank.

Have pupils simulate these conditions. Tell them to half-fill a beaker with syrup or molasses. Add a layer (10-20 ml) of ethyl or isoprophyl alcohol to the syrup. Place the beaker outdoors in an area likely to attract curious insects. Have students check their beakers every day and make written observations.

Insects will inspect the contents, fall into the beaker, and drown in the alcohol bath. Alcohol attracts fruit flies, especially in the spring. A yeast suspension will also draw fruit flies. The alcohol evaporates, brings the insects closer to the syrup which traps them. (Figure 35)

Encourage students to read and report on the tar pits.

Figure 35

ACTIVITY 82

Emphasize few fossils are found in metamorphic rock since tremendous heat and pressure flattens, squeezes, or stretches them almost beyond recognition. Although these great forces cannot be duplicated in the classroom, the following student activity demonstrates the powerful forces of heat and pressure.

1. Fill a beaker half-full of table sugar.
2. Insert a small clam or snail shell in the sugar. Make sure to note the exact position of the shell. Do this by marking the outside of the beaker.
3. Add from 20-30 ml of concentrated sulfuric acid to the sugar. (The instructor should pour the acid.) The acid reacts with the sugar and turns it a dark yellowish brown. Gas bubbles send smoke curling toward the ceiling. A long, cylindrical carbon snake reaches up from the beaker and climbs several inches into the air.
4. When the reaction subsides and cools, carefully scrape away the carbon mass until the specimen becomes exposed. Compare its location and position with the mark on the outside of the beaker (Figure 36). What do you conclude from this experiment?

ACTIVITY 83

How To Rebuild The Past

Mention how geologists dig and sift through tons of rock in search of ancient plant and animal remains. Their reward may be a pile of bones, several broken shells, or a huge dinosaur footprint. If a geologist stumbles over strange bones, he must identify the animal and make inferences based on past discoveries.

Figure 36

Pass out broken bone or shell pieces. Have students replace the missing parts with plaster of Paris or plasticene clay. Provide diagrams, completed skeletons, or text references to aid pupils in their reconstructions.

Suggested Problems And Questions

1. Many groups of animals are missing from the geologic record. Few soft-bodied organisms become fossilized. Tell pupils to half-fill a container (cigar box, tray, etc.) with garden soil. Scatter dead insects, spiders, earthworms, plant twigs, fish bones and shells over the surface. Sketch and label the position of each organism. Cover with one-half inch of dirt. Place the container outside under a bush, pile of rocks, or weedy area. After 5 to 7 days return container to class, remove top dirt layer, and report the results. Students should see that soft-bodied structures disintegrate and decay quickly.
2. Ask students to report on James Hutton's Law of Uniformitarianism. This law suggests the geologic processes which occurred during ancient times (erosion, weathering, etc.) are going on today. Hutton's law proposes that the present is the key to the past.
3. The age of fossil bearing strata can be determined by radioactive

dating. Have pupils report on several methods: (a) Potassium-Argon; (b) Carbon-14; (c) Lead-Uranium-Thorium.

4. Set up a "fossil" dig. Add soil, twigs, leaves, snail shells, clam shells, and fish bones with water. Mix thoroughly. Pour in shallow trays or cardboard containers. Let dry for 2 or 3 days. Pass out containers with these instructions:

   a. Scrape off soil particles. Examine under a microscope or magnifying glass. Sketch and describe what you see.

   b. Record the position—upside down, sideways, etc. and depth (mm) of each fossil specimen.

   c. What inferences can you make concerning type and condition of each fossil? What were the environmental conditions at time of fossilization?

   Encourage pupils to use their imagination.

5. Younger fossils generally appear above older fossils in a strata column. This is known as law of superposition. Hand out clay, and red and white beans. Tell students the red beans are old fossils; the white beans represent young fossils. Have them make models showing these situations:

   a. Young and old fossils appear in the same strata.

   b. Old fossils lie above the young fossils.

   c. Young fossils rest above the old fossils.

   Ask pupils to sketch and interpret these models.

6. Geologists find they can divide rock layers into small units by examining certain fossils. Fossils which help scientists match rocks are called index fossils. Have students list and describe several index fossils.

7. Have pupils report on geologic time divisions: Period, era and epoch. How do each differ?

8. Pass out envelopes containing 4 colored paper strips, 6 inches long by 2 inches wide. The colored strips represent 4 geologic eras: Cenozoic (brown); Mesozoic (green); Paleozoic (blue); and Proterozoic (red). Also, include a chart (Figure 37) listing 14 organisms and their representative symbol. Give pupils a geologic timetable or geology book showing the distribution of fossils. Have them write the letter symbols of organisms on the colored paper which represents the corresponding geologic era (s) and paste on an 8 1/2 inch by 11 inch piece of paper.

INTERPRETING TOPOGRAPHIC MAPS: MAKING YOUR OWN WATER MAP

Introduce the map unit by: (a) listing the terms you choose to stress on the board, then (b) sketch an example of each on an

Symbol	Organism	Time Distribution (eras)*
Cr	Crinoids	Paleozoic to Cenozoic
R	Reptiles	Paleozoic to Cenozoic
Amp.	Amphibians	Paleozoic to Cenozoic
B	Birds	Mesozoic to Cenozoic
Wm	Woolly Mammoths	Cenozoic
T	Trilobites	Archeozoic to Paleozoic
Br	Brachiopods	Proterozoic to Cenozoic
STC	Saber-Tooth Cats	Cenozoic
Ich	Ichthyosaurs	Mesozoic
Bc	Bacteria	Archeozoic to Cenozoic
A	Ammonites	Mesozoic
F	Ferns	Paleozoic to Cenozoic
Fh	Fishes	Paleozoic to Cenozoic
In	Insects	Paleozoic to Cenozoic

Figure 37

overhead projector. For example, the top profile of a mountain can be shown with depression contours, twisting contour lines, elevation markings, streams, roads, building, vegetation, lakes, etc.

ACTIVITY 84

Pass out city, county or state maps. Have pupils find familiar places, e.g., streets, buildings, highways, railroads, parks, rivers, etc. Ask them to give the distances between various points, length (miles) and direction of several roads, plus other outstanding features. Familiar places on local maps make map reading interesting.

ACTIVITY 85

Hand out paper, felt pens, colored pencils, or paints and brushes. Have students make a topographic map which includes:

1. Title, scale and contour interval
2. An area 6 miles wide and 9 miles long
3. Use the scale, 1 inch equals 1 mile
4. 10 foot contour interval
5. 6 roads and 5 buildings (black)
6. 4 lakes, 6 streams and 2 swamps (blue)
7. A woodland or forested area (green)

---

* Let students find these distributions from reference materials.

8. 5 boundary lines (red)
9. 1 railroad running in an east-west direction
10. 3 intermittent streams
11. 2 depressions
12. A steep east side; a sloping west side
13. Highest elevation—180 feet; lowest elevation—60 feet

ACTIVITY 86

Let students make water maps. Provide the following items:

* Large metal cake pans
* White bond paper, 17 inches long by 12 inches wide
* Water
* Oil based paints
* Stirring rod or sticks
* Colored marking pens
* Rulers
* Elmer's glue

Have them fill the tray one-half to three-fourths full of water and add several drops of Elmer's glue. This thickens the water, allowing the paint to spread around easier. Stir the glue and water for approximately 30 seconds.

Add several drops of paint to the water. Each student uses discretion in color choice; he selects the color he feels will produce a desired pattern; i.e., a green-blue hue creates mountains, valleys, plateaus, plains and its neighboring hydrosphere; and brown-green combines various landforms and forested areas.

Tell pupils to hold the paper at both ends and slowly lower it into the tray. When the paper touches the water, hesitate for three seconds. Then, with an upward thrust, bring the paper away from the tray without tipping the ends. Tilting encourages the paint to run across the paper.

Drying time varies. Thick paint takes much longer and creates unwanted smudges. These clumps develop "scars" which damage the final copy. If quick drying is desired, place the map near a heater or slightly above a hot plate. Adding heat causes the paper to buckle and warp, suggesting major diastrophic events; e.g., folding and faulting.

When the map dries, tell the students to carry their papers to their desks or lab tables, set them down, and carefully scan the terrain searching for landform and water features (Figure 38). Encourage pupils to draw contour lines, name volcanic structures, islands, sandbars, glaciers, and bays wherever they occur.

Figure 38

Students need some direction in completing their maps. The following suggestions offer valuable guidelines:

1. Develop a scale-ratio of land distance to map distance.
2. Include symbols for artificial features such as bridges, railroads, buildings, roads, etc., and name each major landform and water area.
3. Label contour lines at regular intervals. Also, compute the total square miles of your map. Hint: Use your scale to calculate miles.

4. Write a brief description of your topographic section. Tell if it is mountainous, flat, heavily forested, etc.
5. Give a brief geologic history of your area. For example, you might explain how the mountains developed in the north or how the sea receded, exposing the present sand bars along the eastern shore.

Suggested Problems And Questions

Offer students these additional items:

1. Review the significance of latitude and longitude lines. Have pupils find the following approximate locations:
   a. Latitude of 22 S, a longitude of 165 °E (New Caledonia)
   b. Latitude of 27°N, a longitude of 82°W (Florida)
   c. Latitude of 42°S, a longitude of 147°E (Tasmania)
   d. Latitude of 68°N, a longitude of 17°W (Iceland)
   e. Latitude of 33°S, a longitude of 57°W (Uruguay)
   f. Latitude of 43°N, a longitude of 155°W (Idaho)
2. Find the approximate latitude and longitude of these places:
   a. Honolulu, Hawaii—(a latitude of 18°N, a longitude of 158°W)
   b. Tokyo, Japan—(A latitude of 36°N, a longitude of 129°E)
   c. Quito, Ecuador—(a latitude of 2°S, a longitude of 78°W)
   d. Manila, Philippines—(a latitude of 14°N, a longitude of 122°E)
   e. New York—(a latitude of 43°N, a longitude of 74°W)
   f. San Francisco—(a latitude of 41°N, a longitude of 124°W)
3. Pass out clay, dissecting needles, nails and wooden splints. Tell pupils to mold an island which includes the following:
   a. Contour interval—2 cm
   b. Highest elevation—6.5 cm
   c. Island length—9 cm
   d. Island width—4 to 6 cm
   e. A steep west side; a sloping east side
   f. 2 depression contours at an elevation of 4 cm
   g. 2 inlets or bays at the north end of the island
4. Gather pieces of plywood or pine. Pass them out and have pupils construct topographic maps, using the grain pattern for contour lines. Encourage students to color their boards and use standard map symbols. The finished products make excellent displays.
5. Demonstrate on an overhead projector how to make topographic profiles. Use a piece of paper, ruled with horizontal lines, to show how perpendicular lines intersect the profile line down to the corresponding horizontal level (Figure 39). Give students sections of topographic maps and have them make their own profiles.

6. How are differences in steepness of slope shown? A series of closed contours usually suggests steepness; widespread contours usually represent a gradual sloping.
7. What do contour lines indicate? Contour lines indicate elevation and shape of the terrain.
8. What are hachured lines and bench marks? Hachure lines are tiny lines drawn perpendicular to contour lines toward center of closed depression. Bench marks are surface spot elevations precisely measured.

Figure 39

SIX SUGGESTIONS FOR FURTHER STUDY

Here are additional activities to assign interested pupils:

1. Erosion is the gradual wearing away of the earth's crust due to water, wind and gravitational forces. Ask students to survey their neighborhood and record evidence of erosional damage. For example, a neighbor's backyard or empty field.
2. Large pieces of the earth's crust break down into smaller substances. This is known as mechanical weathering. No chemical change takes place.

    Test frost action on several porous rocks. Soak different size sandstone rocks in water overnight. Remove and place in freezer. After several hours remove the samples and allow them to thaw. Do any rocks lose grain particles? Repeat several times. Does repeated freezing and thawing show evidence of mechanical weathering?

Experiment to see which of the following items produce the greatest mechanical weathering:

a. Freezing and thawing large porous rock samples.
b. Freezing and thawing small porous rock samples.
c. Extended periods (several days) of freezing and thawing.
d. Short periods (3-5 hours) of freezing and thawing.
e. Type of liquid—distilled water, vinegar, dilute saline or tap water.
f. Rapid freezing, slow thawing; slow freezing, rapid thawing.
g. Type of sedimentary rock—limestone, sandstone, shale, conglomerate, etc.

3. Chemical weathering occurs when certain minerals combine with water, acid, or gas (oxygen and carbon dioxide) to form new products.

   Chemical and mechanical weathering usually occur at the same time.

   Have students test the chemical effect of vinegar, soda water, salt water and dilute hydrochloric acid on limestone, sandstone, and shale chips. Which combination produces a chemical change?

4. Take a field trip to a nearby rocky area. Look for these signs of weathering: exfoliation (peeling rock layers); cracked rocks; loose powder or grains accumulating near rocks (talus); and worn rock edges.

5. Collect fossils in available sedimentary rocks. Examine each layer carefully. Wrap each specimen in tissue or newspaper, give the fossil a number, record the location, return to the laboratory and clean each fossil. Use reference books to identify them.

6. Pour plaster of Paris over a shoe box containing rocks of different size and shape. When the plaster dries, tell students to remove the cardboard and carve a mountain landscape. Have them paint lakes, streams and forested regions on the model.

chapter six

15 ACTIVITIES THAT
TEACH OCEANOGRAPHY

OVERVIEW:

Students investigate saline properties of water, determine percentage salt solutions, and study how the ocean becomes salty.

They examine sonar readings, make sea floor models, and calculate the shape of ocean bottom profiles from these models.

Pupils experiment with wave action and currents. They make plankton from glass beads, string, clay and wire; prepare an algae banquet; and dissect teacher-made "core samples."

Students complete the unit by making paper Spanish galleons and determine which factors caused them to sink.

TESTING THE PROPERTIES OF WATER

Begin this unit by asking students to tell what they know about water. Some obvious statements might be, "It's wet!" "You can drink it!," or "It makes up the rain." These are good starting points. Keep the discussion going by listing these comments on the board and mention the following points:

a. Water has a powerful capacity for storing heat.
b. Most liquids contract or shrink when they solidify, but water expands by 9 per cent when it freezes.
c. When water freezes its volume increases while its mass remains the same.

ACTIVITY 87

Prepare several cubes of different colored ice. Add food coloring or stain to water, stir, and freeze. Pour enough weight (lead shot, iron filings, etc.) into several cubes to make the ice sink. Add the weight when the ice is partially frozen. This prevents it from settling on the bottom. Have the students test the floatability of the ice by placing different colored cubes in a large beaker or tray of water.

Suggested Problems and Questions

1. Ask students to explain why some ice cubes sink and others float. Suggest ways they might test their theories, i.e., weigh the cubes, crush and examine contents, etc.
2. If a student mentions melting the ice cube, ask for an effective way to do this.

Demonstrations And Experiments

1. Prepare ice cubes with 3 different colored layers. Freeze the water one layer at a time. Add dirt, sand, or small pebbles to some of the cubes. Have students repeat the same procedure as in Activity 87. Discuss the results with your class.
2. What factors influence the submergent level of ice cubes? Give students a regular ice cube. Ask them to measure (mm) and record the portion of ice that floats above the water. Make sure each pupil receives a similar size piece. Have them repeat the process with a colored ice cube packed with cork shavings. Do both the ice cubes float at the same level?

Water is often referred to as the universal solvent since it dissolves more substances than any other known liquid. Every living thing—plants and animals—needs water in order to carry on life processes. The sea ultimately provides plants and animals with the water they need.

ACTIVITY 88

Ask the students to compare the water content of several vegetables. Have them bring celery, string beans, radishes, and cucumbers to class. They should cut vegetable pieces of equal size, weigh out 10 grams of each, and place them in 4 different 50 ml beakers. Tell them to drive out the moisture by setting each beaker on a 400° F hot plate (preheated) for 15 to 20 minutes. Advise students to stir the pieces around and try to dry as thoroughly as possible. When most of the moisture has disappeared ask students to remove their beakers from the hot plate, cool the contents, and reweigh them. Have them calculate the percentage of water each vegetable contained. Discuss the possibilities of error with your students.

Students can prepare a water bath by placing a 50 ml beaker inside a 100 ml beaker containing a fair amount of water (40 to 60 ml). Vegetable pieces can be dried out by placing them in the smaller dry beaker and heating the larger beaker of water. This method may be more accurate but takes much longer to complete. *Caution:* Don't use old beakers which might have small cracks in them.

Suggested Problems And Questions

1. Ask students to suggest reasons why some organisms contain more water than others. They should mention factors such as environment, climate, genetic inheritance, evolution, etc.
2. Tell students that some chemical crystals contain various amounts of moisture. This water is known as water of hydration or water of crystallization. Have students suggest ways of extracting this water.
3. Give students 5 dry test tubes and ask them to heat several crystals of copper sulfate, barium chloride, magnesium sulfate, sodium carbonate, and potassium dichromate. Is there evidence of water of hydration? Cobalt chloride paper can be used to determine the presence of water. Place a piece of this paper in the tube. If the cobalt chloride paper turns red, there is water present. You can make your own cobalt chloride paper by mixing 70 grams of cobalt chloride with 100 cc's of water. Soak strips of paper toweling in the solution, dry, and test.

ACTIVITY 89

Give each group of students 4 small plastic bottles numbered one through four. Each bottle should contain one gram of a different substance—ammonium chloride, calcium carbonate, calcium chlorate, and copper acetate. What factors influence the rate of solution? Suggest to the students they might try shaking or heating the test tubes. Ask them to guess which chemicals, if any, are insoluble or soluble in water. Write their responses on the board. Then have them test each chemical. Award a prize (points, etc.) to the winners.

Demonstrations And Experiments

1. You can demonstrate how temperature influences the solubility rate of potassium nitrate. Add .5 grams (1/2 g.) of KNO_3 to 15 ml of water in a Pyrex test tube. Keep adding small amounts of the chemical until a saturated solution is produced. Keep the test

tube in motion by shaking it gently. There should be about 1/4 inch of solute undissolved. Carefully heat the test tube and continue adding more chemical. After the potassium nitrate dissolves, cool the test tube by shaking it under the faucet. The extra solute reappears in the tube.

2. If you live near the ocean, obtain the remains of a dead jellyfish. Ask students for ways of determining the water content of the animal. Give them pieces to test.

WAYS TO TEST SALINITY CONCENTRATION

Almost everyone has strolled down the beach and tasted the salt particles which cling to sea breezes. Two questions come to mind: How salty is the sea and where does this salt come from?

Ask students how salty they think the sea is (percentage salt). Write their answers on the blackboard. Bring up the following ideas:

* The average salinity of sea water is 3.5 per cent.
* Weather forces pound away at mountains gradually wearing them down.
* Minerals trapped within this soil wash into rivers, travel to the sea, and become deposited into the ocean.

ACTIVITY 90

Prepare the following salt solution and ask students to test each solution and decide which one comes the closest to the 3.5 per cent salinity concentration of sea water.

Take 3 large containers—A, B, and C—and pour 1,000 ml's of water into each one. Add 35 grams of sodium chloride to Beaker A, 70 grams to Beaker B, and 120 grams to Beaker C. Stir thoroughly. Give students 10 ml's of each solution to test. Tell students to pour contents into small beaker. They can obtain the weight of the solution by:

* Subtracting the weight of the beaker.
* Boiling away the water.
* Weighing the beaker and remaining material.
* Finding the weight of material by subtracting the weight of the beaker.
* Dividing the weight of water into the weight of remaining material.

*   Multiplying by 100 to receive salinity percentage.

Suggested Problems And Questions

1. Do temperature changes affect the salinity concentration of salt water? Divide the class into 4 groups. Prepare 100 ml's of a 5 per cent salt solution. Give each group a beaker containing 20 ml's of solution. Ask them to set up the following conditions:

        Group 1 —Pack wet paper towels around beaker. Let stand for 10 minutes.

        Group 2 —Place a large ice cube in beaker. Stir around for 10 minutes.

        Group 3 —Heat beaker (slowly) over burner for 10 minutes.

        Group 4 —Place beaker under a heat lamp for 10 minutes.

        After 10 minutes have students compute salinity percentage. There should be no change for Group 1, a slight decrease in salinity for Group 2, and a salinity increase in Group 3 and 4. Heat causes water to evaporate leaving salt particles behind; melting ice adds more water to the beaker thus lessoning the salt concentration.

        Discuss these points with your class using a world map or globe. Mention the sea receives most heat near the equator; the average temperature in the polar regions is approximately $28°$ F. The freezing temperatures of sea water varies with its saltiness, but it is always below $32°$ F. Tell students currents may affect the surface salinities by transporting relatively salty or dilute waters. Surface salinity can be increased by evaporation; can be decreased by melting ice, precipitation, and runoff from the land.

2. Ask students for ways of testing the idea that sea ice in higher latitudes increase the salinity of water. They may not know the answer and wonder how this happens. Tell them when the temperature of sea water is lowered to the freezing point, ice begins to form. Ice tends to exclude the salts and increase the salt concentration in the surrounding unfrozen sea water. Have students test this fact by setting up 4 different beakers with 2, 3, 4 and 8 per cent salt solutions. Place beakers and solution in freezer overnight. Remove and when the solution begins to thaw students can test the salinity percentage in each beaker.

3. What effect would salinity have on a fresh water animal? Students can set up a small aquarium containing a gold fish. Each day they add 40 grams of salt to the tank (fill the tank with 4,000 liters of water). Use the following observation chart as a guide:

Day	Amount of Salt Increase Including Percentage	Breathing— Gill Movement (Beats Per Minute)	Body Movements— Swimming Speed, Mouth and Fin Vibrations, etc.	Overall Activity
1	40 gms (1%)			
2	80 gms (2%)			
3	120 gms (3%)			
4	160 gms (4%)			
5	200 gms (5%)			
6	240 gms (6%)			

ACTIVITY 91

Let students have fun making their own mineral mountain. Have them mix approximately 300 grams of sand, dirt, and clay with 20 grams of sodium chloride. Mold a gently sloping mountain in a dry tray. Fill the tray bottom with fresh water. Simulate rain by liberally sprinkling water over the entire mountain top. Have pupils test the salinity of the tray water after several pourings. Repeat, increasing the pouring time with each trial.

Suggested Problems And Questions

1. Ask your students to check the salinity content of the tray water. Is the salinity reading around 3.5 percent?
2. Suggest that several students calculate the percentage of sediment or debris that washes into the tray. Which soil particles settle first near "shore"—clay, sand, or dirt?
3. Have some students place a large heat lamp over a model mountain for 30 minutes. Then remove the lamp and sprinkle fresh water over the mountain. Does heating cause more salt to enter the tray water?
4. Repeat the experiment. This time remove the heat lamp, insert ice cubes around the sides of the mountain, and pour water around it. Test the tray water.

DOES GRANITE "FLOAT" OVER BASALT? A STUDENT INVESTIGATION

Ask students if they think one rock could "float" over another. This will undoubtedly provoke blank stares. Demonstrate this idea by sliding a piece of granite over a piece of basalt. Mention some scientists feel originally the earth was one large land mass. This single mass, composed mainly of granite rock, slowly drifted apart and "floated" over heavier underlying basaltic material. The ocean floor is composed primarily of basalt. This is known as the Continental Drift Theory.

Students may question that great quantities of rock could be moved in this manner. Tell them some investigators believe convection currents deep in the earth's interior create such a force (See Demonstrations and Experiments). Radioactive elements such as uranium disintegrate, release heat, and push upward toward the earth's surface.

ACTIVITY 92

Have students compare the specific gravity of granite and basalt. Specific gravity is a number that tells how many times heavier a given volume of the substance is than an equal volume of water.

Each student can determine the specific gravity of a substance by:

* Weighing each rock sample on an accurate balance.
* Suspend sample from the balance by a piece of thread and weigh again while completely submerged in a beaker of water (Figure 40).

Figure 40

* Avoid resting rock against the side of the beaker. Subtract the weight in water from the dry weight. The difference is the weight of a volume of water equal to the volume of the mineral.
* Divide the weight of an equal volume of water into the dry weight (weight in air). The final product is the specific gravity.

Suggested Problems And Questions

1. Does shape affect specific gravity? Give students different size pieces of granite and basalt to test.
2. Ask students to list the experimental factors which may cause errors in calculation. Discuss these problems with them.

Demonstrations And Experiments

1. Place pieces of cork, sawdust, or Styrofoam in a three-quarter beaker full of water. Bring the water to a boil. Convective heating causes material to move in a circular pattern. This gives the class an idea of what is meant by convection currents.
2. Give each group of students a box containing cardboard or wood pieces resembling continents. Discuss with them which continents seem to go together.

METHODS OF STUDYING THE OCEAN FLOOR

The idea of sending waves through water is relatively new. Students are already familiar with "echo sounding" or sonar and how these underwater waves were used in World War II to detect enemy submarines. War stories jammed with action do much to stimulate interest and promote classroom discussion. You can relay these stories to your pupils and mention that naval ships and fishing boats alike are equipped with a transmitter for sending out sound waves and a receiver for picking up the returning waves. Tell students on several occasions a school of tuna or a dense water layer was mistaken for enemy craft. Sketch these stories on the blackboard.

Tell students sound waves travel approximately 4,700 feet per second through water. From this figure water depth can easily be calculated. The farther away an object is, the longer it takes the sound wave to reach the receiver. The total sounding time divided in half gives the depth of the object underwater.

ACTIVITY 93

Make several sea floor models. Simply, pour plaster of Paris on the bottom of a fish tank. Add water and mold an ocean bottom. Include deep trenches, high mountains, smooth plains, and sharp slopes. After the mixture dries, fill the tank with muddy water. This will prevent students from seeing the contour of the ocean floor. Have them take from 10 to 15 probes of the bottom while moving from left to right. Each probe should be recorded. For example, Probe 1 equals 8 inches, Probe 2 equals 10 inches, etc. Note: Students can use a ruler to make probes.

Assume 1 inch equals 1,000 feet. Ask students to graph results by placing *ocean depth* on the vertical axis and *trial probes* on the horizontal axis. Drain the tank and allow students to check the accuracy of their results.

Suggested Problems And Questions

1. Have pupils repeat Activity 93 several times. Do the depth readings change each time? Discuss with your students how an average reading is taken from several soundings over the same area.
2. Ask students to tie a string around a small rock or fishing weight and repeat Activity 93. Discuss the results with your class. Tell them this method is ineffective in deep water because ocean currents and water pressure keeps the line from going straight to the bottom.

ACTIVITY 94

Bring up the idea sonar readings taken at different times over the same general area are called profiles. They reveal the true nature of the ocean floor (Figure 41). Show using filmstrips, slides, or transparencies, this once thought "featureless" flatland houses plains, deep trenches, plateaus, hilltops, and mountain ranges. Emphasize the ocean floor is covered with pointed or flat topped volcanic structures known as seamounts.

Figure 41

Ask each student to bring a large potato to class. Have them carve out a pointed or flat topped seamount and slice away 1/4 inch horizontal sections starting from the bottom of the potato. They can make an accurate contour map of their model by placing sliced sections, one at a time, on a piece of white paper and tracing around the edges with a pencil.

Suggested Problems And Questions

1. Tell each student to use a contour interval of 10 feet (Scale: 1/4 inch equals 10 feet) and draw a profile of his seamount.
2. Have students examine several maps of the ocean floor. Suggest they locate several seamounts and note their position. Do they follow any particular pattern? Students will observe seamounts scatter in every direction over the ocean bottom.
3. Ask students to repeat Activity 94 using an acorn squash to make a trough or deep trench, or a small carrot to make a jagged mountain peak.

Demonstrations And Experiments

1. Scientists believe flat topped seamounts were once islands exposed to surface wave action. Earth disturbances caused these islands to submerge. Build a small sand island in a glass tray. Add water and show how wave action erodes the island.
2. Give students some plaster of Paris, flour and water, or clay. Have them make their own ocean bottom. Suggest they add dirt, pebbles, or sand as they see fit. Provide water colors to those who wish to paint their models.

INVESTIGATING FACTORS WHICH CAUSE OCEAN CURRENTS

Ask students what they think keeps the ocean water moving. Write their answers on the blackboard. Discuss such factors as wind, earth rotation, water density (cold water sinks, warm water rises,) the depth and shape of the ocean bottom; and the contours of countries touching the sea. Use a world globe to help demonstrate these points. Mention that volcanic eruptions, earthquakes, and strong winds have been blamed for sending large ocean waves, called tsunamis, crashing into coastal villages causing tremendous damage and great loss of lives.

ACTIVITY 95

Give each pair of students 2 - 250 ml Erlenmeyer flasks or gas collecting bottles and a glass plate. Ask them to:

1. Fill one container with hot, colored water and the other container with cold water (these can easily be prepared in advance).

2. Place the glass plate on one container, turn it over, and set it on top of the other container. Hold both containers in an upright position making sure the mouths of the containers line up properly. Remove the glass plate (Figure 42).
3. Repeat the experiment using salty, colored water and clear, fresh water. Does varying the salt concentration have an effect on water circulation? Students should see by adding salt the water circulation increases.

Figure 42

Suggested Problems And Questions

1. Ask students to repeat Activity 95 using different size bottles and warm instead of hot water. They will observe warm water rising.
2. Have students fill one flask with cold water and another with hot, muddy water. Does sediment in water affect circulation? Pupils should see the slow settling sediment has little effect on water movement. However, fast moving sediment known as turbidity currents are believed to carve out trenches or submarine canyons in the ocean bottom. Discuss these currents with your class.

Tell your students water loss through evaporation not only adds moisture to the atmosphere but increases the saltiness and density of the surface water. This affects water movement.

The Mediterranean sea loses on the average 100,000 tons of water per second due to evaporation. Ask students why the Mediterranean sea does not dry up. Some may say it probably is, but will take a long time; others may hint evaporated sea water returns to the sea. Relate how sea water returns via rain, snow, sleet, or hail. Use available audio-visual guides and charts to explain the hydrologic cycle.

ACTIVITY 96

Have students add 30 grams of salt to 100 grams of water, mix thoroughly, and pour the solution into a 200 ml beaker. Ask them to measure the water depth and calculate the percentage of salt concentration. Boil for 10 minutes. Repeat the above measurement. They should notice a reduced water level and an increase in salinity. A white, salty powder is noticeable around the inside walls of the beaker.

Suggested Problems And Questions

1. Ask students to repeat Activity 95 using different size bottles and warm instead of hot water. They will observe warm water rising. They should realize the sun draws moisture and salt particles into the air.

2. Give students 20 ml's of 30 percent salt solution. Have them boil solution until it disappears. Tell them to add 20 ml's of fresh water, stir, and boil again until no more remains. How many times must they repeat this process until no more salt remains. Does this process occur in nature?

Ask your students what tides are, what causes them, and why fishermen consider them important. Use models to demonstrate the relative positions of the earth, moon, and sun. Explain how the earth and moon exert a mutual gravitational force on each other. The moon's gravitational force causes the periodic rise and fall of the level of the sea. This phenomena is known as the tides. The sun's gravitational force also raises tidal bulges on the earth, but to a lesser degree because of the sun's greater distance from the earth. Fishermen know tides keep water in motion and to some degree influence how the fish bite.

Introduce the idea ocean waves are caused by wind action. Sketch a water wave on the blackboard and label the crest, trough, wavelength, and wave height. Mention waves do not actually move water forward, but individual water particles within the wave move in circles. Demonstrate this by setting a cork in a large tank, produce waves, and observe the cork's motion. It moves forward slightly as the crest of a wave approaches and falls back a short distance as it passes into the trough. It remains in almost its original position.

SIX WAYS TO CREATE OCEAN WAVES

ACTIVITY 97

Have your students make a model ocean. Tell them to fill a rectangular plastic or glass container with approximately one inch of water. They can produce waves by any of the following methods:

1. Roll a metal rod back and forth along the bottom of the container.
2. Bump one end of the container with regular sharp taps of the hand.
3. Set a fan (with adjustable speed) near one of the ends.
4. Blow across the surface of the water.
5. Release water from a medicine dropper, one drop at a time, near one of the ends.
6. Cup one hand, dip it slightly in the water, and slowly move it back and forth.

Have students devise tests to determine wave speed, wave length, and height, and how water depth might affect wave action.

Demonstrations And Experiments

1. You can construct a simple ripple tank by placing a piece of white paper on a ring stand, setting a Pyrex baking dish on top of the paper, adding water to the dish, and hanging a light over the dish. Wave action can be observed on the white paper.
2. Demonstrate wave action by attaching one end of a rope or "Slinky" spring to a solid object. Move the free end in an up and down motion.
3. Have students attach a small cotton bag filled with salad or vegetable oil next to a wooden block (the block represents a boat). Ask them to place the block near the center of the container and make a series of small waves. Discuss the results with your class. Does oil reduce wave action?
4. Fill one end of a container with sand. Build a beach or shallow shore line. Ask students to observe what happens to waves when they reach the sand. Discuss beakers and why surfers find them so popular.

HOW TO MAKE PLANKTON MODELS

Most students realize the sea abounds with life. Use available slides, charts, and filmstrips to discuss the various forms of sea life.

Environmental conditions—obtaining oxygen, how buoyancy supports movement, temperature factors, and plant-animal relationships. Allow students to examine bottled specimens or plastic mounts of sea life. This encourages further discussion and study. Mention the value of food chains and how larger animals such as the blue whale depend upon the presence of plankton, free-floating marine organisms, for its existence. Stories of killer whales, sperm whales, giant squids, and man-eating sharks arouse considerable interest.

ACTIVITY 98

Emphasize how plankton supply larger animals with food. Diatoms (microscopic algae) are members of the plankton family. Together they supply all sea animals with an abundance of food.

Give each student a box containing wire, thread, string, glue, plastic, or glass beads of different shapes, sizes, and colors (Figure 43). Have them make plankton. Provide reference books or transparenices illustrating several varieties of plankton.

Figure 43

Suggested Problems And Questions

1. Model plankton can also be made from modeling clay or Styrofoam packing "worms." Let students put on a plankton carnival. Award prizes for the greatest variety, the most colorful, etc.

2. List several examples of sea life on the blackboard. Include ocean plant specimens. Ask students to construct a food chain.

Inquire whether or not your students have ever eaten snails, chocolate covered ants, french fried grasshoppers, or rattle snake. The typical "Oh, how sickening" response usually prevails. Mention man continually searches for edible sea food products. Sea farming, known as aquaculture, exists in Pacific coastal areas, both in Japan and the United States. Some people claim that plankton tastes like lobster, shrimp, or vegetables. Large algae harvests produce low-calorie diet products. Algae finds it way into cheese, canned mushrooms, ice cream, and malted milk.

ACTIVITY 99

Give students several pieces of algae. Ask them to try different ways of preparing a seaweed dinner. If they have trouble getting started, you might suggest ways to sterilize the plant, remove the salt, and tenderize the plant tissue. Hold a contest. Ask students to bring different spices to class and experiment to see which ones add flavor to their cuisines. Taste each preparation and award a prize for the best recipe. Note: If algae isn't accessible, substitute with water plants. Make sure they're "edible."

HOW TO MAKE A CORE SAMPLE

Discuss the nature of ocean bottom sediment with your students. Bring up how sea bottom sediments differ from one area to the next. Coarse rock fragments and sand are found near shore; fine mud and clay particles dominate the bottom in deeper zones. Volcanic ash, meteoric dust, calcite and silicate shell material settle over the ocean floor. When plankton die, they shower the ocean bottom with a heavy layer of very fine slimy sediment called ooze.

Ask your students how scientists know what lies on the sea bottom. Answers should range from sonar readings to diving bells. Go a step further and mention scientists use core samplers to study the bottom sediments. These samplers are tube-like instruments which dig into the ocean floor and bring up a core which shows the composition of the material for several feet down.

ACTIVITY 100

Prepare a mixture of clay, water, broken sea shells, ash, and sand together. Mold into a soft doughy substance. Fill paper or plastic tubes (3 to 5 inches long) with the mixture (Figure 44). Tell the students these cylinders represent core samplers taken in 300 feet of water. Have them do the following:

a. Carefully slice the tube lengthwise.
b. Sketch and identify each substance found in the mixture. Provide a magnifying glass or microscope to help them with their identification.
c. Describe the type of environment which exists and the kinds of living organisms that live there.

Figure 44

HOW TO SINK A SPANISH GALLEON

Stories stressing sunken Spanish galleons with hulls bulging with gold and silver coin excite nearly every pupil. More questions arise than can be answered. You may successfully end this unit by asking your students to test some of the possible reasons these 15th century galleons never reached port.

ACTIVITY 101

Give each student a 5 by 8 inch piece of heavy bond or Manila folder paper. Add the following instructions:

1. Make any size or shape boat you wish. Use staples, paper clips, scotch or masking tape, or glue to seal the exposed end.
2. Fill a large glass, plastic container, or sink three-quarters full of water.
3. Add small amounts of a known weight to the boat. Continue to add weight until the boat sinks. Make sure students record the total amount of weight necessary to submerge their boats.

Suggested Problems And Questions

1. Offer a prize to the student who can keep his boat afloat the longest. Encourage students to investigate ways they can keep their ships from sinking.
2. Have students repeat Activity 101 and test the following questions:
   a. Will the boat remain afloat longer in fresh or salt water?
   b. Try various concentrations of salt water. Does the salinity of the water seem to make a difference in the boat's ability to stay afloat?
   c. Shift the weight to different positions around the boat. Which area of the boat's bottom appears to hold the most weight?
   d. Place several different shaped boats in the water. Create waves across the surface of the water. Which boat stays afloat the longest? The shortest?
3. Does the shape of the weight influence how long the boat will remain afloat?
4. Does the water depth affect a boat's ability to stay afloat?

When students complete these experiments, discuss with them the factors which may have caused so many galleons to sink.

ELEVEN SUGGESTIONS FOR FURTHER STUDY

A. Assign any of the following experiments to students who wish to investigate further, need to make up work, or desire extra credit.

1. Find out the lowest temperature which can be produced with an ice and salt mixture.
2. Which freezes first—cold or hot water? Try an equal volume of each.
3. Find the water content of fresh carrots, beets, watermelon, and cantaloupe.
4. Which contains the most moisture—fresh, frozen, or canned corn kernels?

5. Compare the water content of a cactus plant and a water lily.
6. The following chart lists different minerals and their approximate percentages which can be found in an average sample of sea water.

Mineral	Average Percent
Sodium Chloride (Na Cl)	77
Magnesium Chloride ($MgCl_2$)	12
Magnesium Sulfate ($MgSO_4$)	4
Calcium Sulfate ($CaSO_4$)	3.4
Potassium Sulfate (K_2SO_4)	2.3
Calcium Carbonate ($CaCO_3$)	.3
Magnesium Bromide ($MgBr_2$)	.2

Make a 100 ml "sea water" solution from the information above. Do this by multiplying the *Average* Percent column of *each* mineral by 3.5 per cent. The answer will be the amount of mineral (in grams) to add to 100 ml of water. Since calcium carbonate and magnesium bromide measurements are extremely small, add a tiny pinch of each to the solution.

7. If a sound wave takes 4 seconds to leave a ship—touch a submerged object—and return again, the object is located 9,400 feet below the surface (sound waves travel approximately 4,700 feet per second). Determine the depth of the following 5 objects (divide the *total time* in half).

OBJECT	TOTAL SOUNDING TIME	DEPTH (in feet)
1. Mountain Top	.8 seconds	?
2. Guyot (seamount)	1.7 seconds	?
3. Ocean Bottom	2.4 seconds	?
4. Submarine	.3 seconds	?
5. School of Fish	.4 seconds	?

8. Sketch an ocean profile on a piece of graph paper. Include the following structures:
   a. 3 seamounts
   b. 1 plateau
   c. 1 slope
   d. 2 islands
   e. 1 submarine canyon

Use the scale 1 inch equals 2 miles and one-quarter inch equals 1 fathom (2 fathom equals 6 feet). Place *distance* on the horizontal axis and *depth* on the vertical axis. Write a brief summary of your profile and give the depth and distance for each structure.

9. Using the following mineral chart as a reference, measure the specific gravity for *each* mineral composing granite and basalt. Your teacher will provide these minerals for you. Calculate the average specific gravity for all minerals which make up each rock. Figure the specific gravity for a solid piece of granite and basalt. How do the two figures compare? If they differ, how would you explain the results?

GRANITE

MINERAL	SPECIFIC GRAVITY
Quartz	2.65
Orthoclase Feldspar	2.50 - 2.62
Muscovite (mica)	2.76 - 3.0
Hornblende	3.05 - 3.47

BASALT

MINERAL	SPECIFIC GRAVITY
Plagioclase Feldspar	2.6 - 2.7
Olivine	3.2 - 3.6
Augite	3.19
Magnetite	5.17 - 5.18

10. A small group of students can build a sonar box from 1/4 inch plywood. The directions are as follows:
    * Make 2 sides of the box 19 inches or more wide.
    * Make the other 2 sides 7 inches or more wide.
    * Cut 3 - 1/4 inch grooves into one side of the box. This allows 3 - 6 inch adjustable metal bolts (with a single nut, 2 large, round thin washers, and 1 wing nut) to move up and down. A 3 piece plywood shelf rests against the metal bolts and can be adjusted to different levels which represent various ocean depths (Figure 45).

    Students pair off. One partner bounces a Ping-Pong or super ball at least 3 times against each level. The other person holds a stop watch and times how long the ball takes to pass by the top of the box and return again. He should stay at eye level with the top of the box. Partners take turns timing and bouncing the ball. Suggest that each pupil graph his readings. Assume .2 seconds equal 1,000 feet. Place the *ocean depth* on the vertical axis and *trial bounces* on the horizontal axis. Have the students test these problems:
    1. What variables might influence your readings?

Figure 45

2. How would adjusting the ocean floor affect the speed of the ball?
3. Place different materials—rubber, cloth, Styrofoam, paper—on the various levels. How does the ball bounce off of each material?
11. Set up an experiment to determine how piles of sand, and large rocks or boulders affect wave speed and direction.

UNIT TWO

EXPLORING LIFE

chapter seven

17 ACTIVITIES THAT HELP TO EXPLAIN MICROORGANISMS

OVERVIEW

The section opens with an introduction of the 3 types of bacteria. Student group activities stress bacterial products and conditions under which they live. Pupils observe bacterial growth in soil, meat, milk and vegetables. Experiments include spoilage rate of bacteria and decomposition (putrefaction).

Students spend several days studying Chromobacter violaceum, a purple-pigmented bacterium which can be seen without a microscope. The unit closes with a microbial feast and final observation of prepared investigations.

HOW TO CULTURE MICROORGANISMS FROM GRASS, LEAVES, SOIL, AND WOOD SCRAPS

Begin by asking students why it's important to keep clean, eat foods which are properly cooked, and apply antiseptic to fresh injuries. Most agree harmful germs thrive in filth and can easily cause infection. Ask if all bacteria and germs are dangerous; also, how one might tell a destructive germ from a harmless one. Answers will vary.

Introduce 3 types of bacteria: Larry, Curly and Mo. Tell pupils bacteria are microorganisms which contain no chlorophyll, can be found almost anywhere, and multiply rapidly by simple division. Larry, the cocci, boasts a round body (hold up a ball); Curly, the bacilli, carries a rod-shaped carcass (show a metal cylinder, ruler, etc.); and Mo, the spirilla, displays a snake-like frame (hold up a twisted or coiled spring). These bacteria roam in stooge-like fashion throughout soil, manure piles, milk products, water—everywhere!

They're blamed for causing such problems as pimples, boils, typhoid fever and tuberculosis. However, most bacteria are relatively harmless.

ACTIVITY 102

Take the students on a "slapstick hunt." Have them walk the campus and collect a handful of dried grass, leaves, soil and wood scraps. Upon returning to the class room, pass out 4 - 500 ml beakers to each group of students. Tell them to label each beaker, e.g., group number, beaker letter (A, B, C, D), contents and date (Figure 46). Have pupils prepare beakers as follows:

Beaker Letter	*Contents*
A	Handful of dried grass, 2/3 full of distilled water.
B	Handful of dried leaves, 2/3 full of distilled water.
C	Handful of dry soil, 2/3 full of distilled water.
D	Handful of wood, cloth or paper scraps, 2/3 full of distilled water.

Figure 46

Stir ingredients for approximately 30 seconds. Let mixtures stand for a few days until a scum forms on the surface. Have students make several microscopic slides from each beaker and sketch what they see under high power. Suggest pupils prepare several stained microscopic slides ahead of time. Tell them to add a drop of methylene blue to each slide, allow to dry, and set aside until needed. When a drop or smear containing bacteria is added to the stain, the methylene will make the bacteria easier to see. Mention how moving bacteria show up clearer than stationary organisms. Tell students to check beakers every 2 days for 6 days. Beakers should be kept in warm area.

The first day observation probably will reveal few organisms. Assign the following questions for the remaining observations:

1. Does every beaker contain bacteria? If not, why not? Answers will vary.
2. What conditions encourage bacterial growth? Students might say moisture, food supply, temperature, absence of contaminants, type and shape of container, etc.
3. What conditions retard bacterial growth? Answers might include adverse temperatures, contamination and reduced food supply.
4. Are the 3 types of bacteria present in each beaker? Answers will vary.
5. Are other organisms present? Pupils may see strange microorganisms and molds develop.
6. Why might molds and pathogenic organisms appear in the beakers? Bring up the importance of sterilization. Tell students unsterilized equipment opens the door for possible germ invasion.

Provide each student group with a large container of Lysol solution. Pupils can rinse off their equipment and store it overnight in the Lysol bath. Remind them to wash their hands before leaving the room.

Emphasize bacteria swim in water; float through the air; stick to soil particles; accompany sneezes, coughs, and handshakes; and hide in unsuspecting places. Tell students bacteria may be present even though they cannot be seen. And every microorganism should be treated as a pathogenic agent.

RAISING BACTERIA FROM BEANS

ACTIVITY 103

Have students soak white, pink, pinto, or lima beans in water for approximately 3 days. Then extract liquid around beans with a clean eye dropper. Prepare several microscopic slides and observe bacteria under high power. Record your findings.

HOW TO GROW A BACTERIAL SMORGASBORD

ACTIVITY 104

Let pupils set up a bacterial smorgasbord. Give each group 4 test tubes and a container or stand to hold the test tubes. Tell them to label materials (See Activity 102). Have pupils prepare tubes as follows:

Test Tube	*Contents*
A	A small piece of liver, 30 ml of distilled water.
B	A small piece of hamburger, 30 ml , of distilled water.
C	A potato slice, 30 ml of distilled water.
D	A tomato chunk, 30 ml of distilled water.

Insert a cotton plug in each test tube. Gently shake the contents. Let stand for several days. Keep in dark, warm place. Have students make the second observation of Activity 102.

Suggested Problems And Questions

1. Do all beans produce the same type of bacteria? Have pupils isolate and soak different kinds of beans in water for several days. Extract and examine the liquid from each sample.
2. Tell students to place the following products in distilled water and soak for several days:
    a. Fresh kernel corn
    b. Frozen kernel corn (uncooked)
    c. Frozen kernel corn (cooked)
    d. Canned kernel corn (uncooked)
    e. Canned kernel corn (cooked)
    Which medium produced the largest amount of bacteria?
3. Have them repeat Problem 2. Use mashed, boiled (unmashed), fried and raw potatoes.
4. Can bacteria be easily seen? Instruct pupils to rub their fingers against the chalk board, across a laboratory sink, or along a window sill. Then transfer material on a microscopic slide and

observe under high power. Generally, few bacteria will be found in dry places. Remind students to wash their hands.
5. Have students set up 8 containers. Place 6 or 7 fresh peas in each container. Add 40 - 50 ml of the following solutions to the different containers: (a) 2% salt, (b) 5% salt, (c) 15% salt, (d) 2% sugar, (e) 5% sugar, (f) 25% sugar, (g) vinegar, (h) distilled water. Which solutions produce the largest bacterial population?

HOW TO OBSERVE MILK SPOILAGE

Relate how fast growing bacteria causes chemical and physical changes to occur in certain food products. This condition is known as spoilage. Bread mold growth and sour milk are excellent examples. These conditions help destroy bacterial growth: (a) application of extreme cold or hot temperatures, (b) reduced oxygen supply, (c) absence of moisture, (d) drying products, (e) increased acidity, (f) salting.

ACTIVITY 105

At what rate does milk spoil? Give each student group 3 beakers. Have pupils pour about 30 ml of fresh milk into each container and set them in a different location, e.g., refrigerator, dark cabinet, near a heater, window, or light source. Label each beaker. Make a daily check for 7 to 10 days. Prepare microscopic slides and transfer milk to the slides by a pipette or eye dropper. Record findings. Include evidence of spoilage—curdling, strong odor, color changes, etc. Caution students not to taste any materials.

Discuss Pasteurization, the process for killing pathogenic germs and bacteria in milk, beverages and other food substances. Pasteur found milk heated to approximately 145°F for 30 minutes, then rapidly cooled to 50°F, would destroy contaminants. Mention Pasteurization doesn't kill all bacteria. Some live to turn milk sour.

ACTIVITY 106

Obtain a quantity of raw, unsterilized milk. Have student groups "Pasteurize" about 40 ml of raw milk by heating (145°F), then cooling (50°F). Place in a sterilized beaker. Pour the same amount of untreated raw milk in an unsterilized beaker. Set aside. For the next few days check the spoiling rate of each beaker. Label all containers. Have pupils examine and record their findings for Activity 103 (bean bacteria) and Activity 105 (milk spoilage).

ACTIVITY 107

Tell pupils to check the bacterial growth of Activities 102, 103, 104, 105 and 106. Remind students to make sketches and careful recordings for each investigation; also, to clean all equipment and store in the Lysol solution.

Suggested Problems And Questions

1. Have pupils test the spoilage rate of different kinds of fruit juice, e.g., pineapple, orange, apple, grapefruit and prune.
2. Repeat Problem 1. Test the spoilage rate of open and closed containers.
3. Give students two thin pieces of bacon. Have them completely cover one strip with salt and leave the other untreated. Which strip cultivates the most bacteria or mold?
4. Repeat Problem 3. How does beef jerky, salted bacon and untreated bacon compare?
5. Have pupils compare the spoilage rate of fresh milk, canned milk and powdered milk.
6. Ask students to set up displays which show how bacterial growth can be reduced. Let them experiment with different solid and liquid products. Note: Let pupils find out how slowly milk spoils at 40, 36, 34 degrees Fahrenheit.

STUDYING NITROGEN-FIXING BACTERIA

Discuss how bacteria helps man improve his environment and adds quality to world products. Describe with the aid of diagrams, filmstrips or slides, the role bacteria play in the carbon and nitrogen cycle. Emphasize the following points:
1. Bacteria play an important role in making nitrogen available for re-use.
2. Bacterial action breaks down carbon compounds into carbon dioxide.
3. Bacteria gives cheese an excellent texture, aroma and taste. Cheddar and Limburger cheese are good examples.
4. Bacteria aid in curing leather and tobacco.
5. Acetic acid bacteria hastens the fermentation of alcoholic products.
6. Bacteria help break down waste products changing solid materials to gas.
7. Bacteria aid in putrefaction or decomposing organic matter.

ACTIVITY 108

Mention nitrogen-fixation. Tell students certain soil bacteria enter the roots of leguminous plants (clover, peas, beans and peanuts), feed off the plant juices, reproduce, build nodules or lumps on the roots, remove nitrogen from the air, and provide nitrates for the plant.

Take students on a short field trip around the school. Look for clover plants with root nodules. Bring several back to the classroom. Tell pupils to prepare several methylene blue stained slides. After they dry, press a nodule between 2 glass slides, crush, and remove one of the slides. Transfer juice from the nodule (toothpick or needle) to a stained slide, add a drop of water, cover slip and observe under high power.

Check Activities 103, 105 and 106 for bacterial growth.

INVESTIGATING PUTREFACTION

ACTIVITY 109

Stress how anaerobic bacteria (organisms which fluorish without free oxygen) break down or chemically decompose organic matter. This is known as putrefaction. Enzymes diffuse out of bacterial cell walls to hasten chemical breakdown.

Give each group of students 5 dead earthworms, 5 beakers, paper, soil, and rubber bands. Have them prepare beakers as follows:

Beaker	*Contents*
Number 1	Fill beaker 2/3 full of moist soil. Lay earthworm on top of soil. Cover with paper and hold down with rubber bands. Place in cool area.
Number 2	Fill beaker 2/3 full of dry soil. Lay earthworm on top of soil. Cover with paper and hold down with rubber bands. Keep at room temperature.
Number 3	Fill beaker 2/3 full of moist soil. Lay earthworm on top of soil. Leave container open. Keep at room temperature.
Number 4	Fill beaker 2/3 full of moist soil. Lay earthworm on top of soil. Cover with paper and hold down with rubber bands. Place outside in warm area.

Number 5   Place an earthworm in an empty beaker. Leave container open. Keep at room temperature.

Check beakers every 2 or 3 days for the remainder of the unit. Record all observations and note evidence of decay in each beaker. Check Activities 103, 105 and 106 for bacterial growth. Set Activity 103 aside.

NINE WAYS TO OBSERVE CHROMOBACTERIUM VIOLACEUM

The remaining activities center on Chromobacterium violaceum, a purple pigmented bacterium which grows easily at room temperature in Nutrient agar, Nutrient broth, or beef bouillon.

A culture of Chromobacter can be obtained from biological supply houses or local universities or colleges. The organism should be stored in the refrigerator and transferred with a sterile loop to a new growth medium once a month. After transfer, the organism should be allowed to grow at room temperature for 48 hours and then be refrigerated. For long-term storage, the organisms may be frozen and brought to room temperature slowly when needed. Chromobacter is non-pathogenic and safe to use in the classroom.*

Culture plates and equipment can be prepared in the following manner:

1. Prepare agar medium (follow the directions found on the container).
2. Pour agar into Petri dishes. Sterilize Petri dishes and lids in a 305°F oven for 2 hours. Package cotton swabs and plugs. Place in oven with Petri dishes.
3. Petri dishes may be sterilized in an auto clave (pressure cooker). Set at 15 pounds pressure for 15 minutes.

The following exercises contain no procedures or lists of instructions. The student receives the problem to be examined. He must work out his own questions, procedures, results, observations, and conclusions.

------

* A non-pathogenic bacterium must still be treated with caution. Remind students to thoroughly wash their hands and all equipment following use.

ACTIVITY 110

Who Is Chromobacter?

Students examine Chromobacter on its solid medium, Nutrient agar, and its liquid growth medium, Nutrient broth. The pupils discover on a solid medium Chromobacter appears as a brightly colored purple circular colony. In a liquid medium, the growth of Chromobacter shows up as cloudiness in the liquid and an accumulation of purple cells at the bottom of the test tube after 48 hours. If students view Chromobacter on a methylene blue stained slide, they will see rod-shaped organisms.

Have pupils check bacterial growth on Activities 104, 105 and 106. Set Activity 104 aside.

ACTIVITY 111

Where Is Chromobacter?

To hunt for Chromobacter and other microorganisms, each student needs a Petri dish containing Nutrient agar and sterile cotton swabs. A few containers of sterile water or Nutrient broth should be available in the classroom. The pupil decides which object he wants to investigate: his fingers, his comb, his shoe, the desk, the drinking fountain. The student then decides whether the object to be investigated is small enough to be applied directly to the Petri dish or if the organisms must first be picked up by a cotton swab, moistened with sterile water or sterile Nutrient broth, and applied to the Petri dish.

After 48 hours smooth colonies of bacteria and the characteristic fuzzy mycelial growth of fungi become apparent. Have students observe the organisms growing on their object of investigation. Encourage them to compare their results with the objects studied by classmates, and search for the purple-colored colonies of Chromobacter.

Have pupils check bacterial growth on Activities 105, 106 and 109.

ACTIVITY 112

How Easily Can Chromobacter Be Spread?

Students "shake hands" with a contaminated candy bar. This experiment shows the ability of Chromobacter to spread purple-pigmented colonies. Groups of 10 students carry out the activity.

Provide one Petri dish, 4 sterile swabs, and one candy bar by inoculating its surface with a sterile swab. The first student "shakes hands" with the candy bar; that is, touches the contaminated area with his fingers. He then touches fingers with the second student and proceeds to touch his fingers to the surface of the Petri dish. The second pupil touches fingers with the third student and then "plates out" his fingers. The group repeats the procedure until all of the students have "exposed," "transmitted," and "plated out" their fingers. Have pupils examine the plates 48 hours later. The number of plates containing purple fingerprints is an indication of how far an organism can be transmitted through direct contact.

Have pupils check bacterial growth on Activities 105, 106 and 110.

ACTIVITY 113

What Are The Minimal Nutrients Chromobacter Needs For Growth?

In this experiment the students prepare a defined growth medium. The recipe for the medium is as follows:

Solution A

Glucose	0.5g
Sodium Chloride	0.5g
Ammonium Monophosphate	0.1g
Potassium Diphosphate	0.1g
Water	90 ml

Solution B (Prepared in advance by instructor)

Magnesium sulfate	0.2g
Ferrous sulfate	0.1g
Calcium Chloride	0.1g
Water	1000 ml

Add ten milliliters of Solution B (inorganic salts) to Solution A.

Divide the class into 4 groups. Group 1 prepares the medium following the given recipe exactly. Group 2 omits glucose, the only carbohydrate and the source of carbon. Group 3 omits Ammonium monophosphate, the only source of nitrogen, an essential constituent of protein. Group 4 prepares the medium omitting the inorganic salt solution and adding instead 100 milliliters of water in Solution A. Prepare the media in 15 Erlenmeyer flasks, plugged with cotton, and sterilized. After sterilization inoculate the flasks with 0.5 milliliters, or a sterile medicine dropperful of a Chromobacter broth culture.

Have pupils observe the flask daily. After 4 days students should see the following results:

1. The complete medium becomes turbid, indicating growth has occurred.
2. Glucose-free medium remains clear.
3. Ammonium monophosphate-free medium stays clear or becomes slightly turbid.
4. Inorganic salt-free medium turns slightly turbid. What has happened? Chromobacter grows in the complete medium because everything necessary for its growth is present. It cannot grow in the medium from which glucose is absent because the glucose is necessary as a carbon and energy source. Nitrogen is necessary for protein synthesis. The magnesium, iron, and calcium contained in the inorganic salt solution are also essential in trace amounts of enzymatic activity to occur within the life processes of the cell. The third and fourth flasks show slight growth because impurities, e.g., nitrates, magnesium, iron, and calcium found in the chemicals offer enough food for limited growth.

Have pupils check bacterial growth on Activities 105, 106, 109 and 111.

ACTIVITY 114

Does All Living Matter Possess Enzymes? Does Pineapple?

Tell students the compounds an organism uses as nutrients and converts to cell products are directly related to the enzymes it possesses. Enzymes, biological catalysts necessary for many reactions, are part of every living organisms.

The following pineapple experiment helps pupils visualize an enzymatic reaction: Divide the class into 3 groups. Emphasize the directions given on the packages of gelatin dessert—"Do not add fresh or frozen pineapple. Use only canned or cooked pineapple." Ask pupils why this directive is necessary. Have Group 1 add canned pineapple to a gelatin package; Group 2 adds fresh pineapple to their gelatin preparation; Group 3 boils the fresh pineapple for 10 minutes and adds this boiled pineapple to their gelatin. After an hour and a half the gelatin containing the canned and the one containing the cooked pineapple jelled, but the one containing the fresh pineapple is completely liquid. Pineapple contains an enzyme which breaks down gelatin. Broken down gelatin does not solidify. Heat causes the enzyme, which is protein in nature, to become inactivated. Thus the gelatin jells.

Have pupils check bacterial growth on Activities 105, 106, 112 and 113.

ACTIVITY 115

How Does Temperature Affect Chromobacter?

Give each group of 4 students 4 test tubes containing sterile Nutrient broth, a sterile medicine dropper, and a culture of Chromobacter. Have them inoculate one drop of culture into each tube, and one tube from each group is incubated in a freezer, in a refrigerator, at room temperature, or in boiling water for 48 hours (the tube to be boiled can be boiled for 10 minutes and then incubated at room temperature). After 48 hours have pupils examine the tubes for growth as evidenced by turbidity.

How can students determine whether organisms are actually "dead" or whether their growth has simply been inhibited? They can reincubate the tubes which showed no sign of growth at the initial incubation temperature at a temperature at which growth did occur. If the organisms are still living, growth will occur in 48 hours.

Have pupils check bacterial growth on Activities 105, 106, 109 and 113.

ACTIVITY 116

Romping the Environment

Let each student design his or her own experiment. Allow pupils to investigate additional environmental factors which may or may not affect Chromobacter growth. Other factors include ultraviolet light, disinfectants, chemicals, detergents, acids, gases, etc. Provide Nutrient agar plates, Nutrient broth tubes, sterile medicine droppers, and cotton swabs for transfer.

Glass jars make suitable Petri dishes. They can be sterilized in a dishwasher or in the oven at 350°F for 2 hours after first removing any cardboard or rubber lining from the lid. These provide adequate containers for agar. A pupil wishing to test a gaseous atmosphere need only inoculate the hardened agar with Chromobacter, introduce the gas, replace the lid, and seal the jar with clay or plasticene.

Have pupils check bacterial growth on Activities 105, 106, 113 and 115.

ACTIVITY 117

Is Chromobacter Sensitive to Penicillin And Other Antibiotics?

Test the effect of known antibiotics upon Chromobacter. Give each student a Nutrient agar plate. Make available sensitivity discs (can be obtained from biological supply houses or clinical laboratories), sterile swabs, Chromobacter cultures, burners, 70% alcohol and tweezers. Tell students to heavily streak the entire surface of the plate with Chromobacter. They can sterilize the tweezers by dipping them in alcohol and running them through the flame. Remove 6 to 8 sensitivity discs from the container with the sterilized tweezers and press gently onto the agar. After 48 hours, areas of no growth can be observed around those discs containing antibiotics which have the ability to inhibit the growth of Chromobacter. The antibiotics which cannot inhibit Chromobacter will have growth all around the disc.

Have pupils check bacterial growth on Activities 105, 106, 113 and 116.

ACTIVITY 118

Microbial Feast

End the unit on a happy note. Hold a party! Have pupils bring to class microbial products, e.g., pizza, bread, cheese, yogurt, pickles, sauerkraut, etc.

Finish observing bacterial growth on Activities 105, 106, 109, 113, 116, and 117.

TEN SUGGESTIONS FOR FURTHER STUDY

Assign the following problems to interested students:

1. Make stained slides of different flavored yogurt, e.g., strawberry, peach, raspberry or blueberry. Do they all contain bacteria?
2. Make a microscopic investigation of sauerkraut juice. Do you see evidence of bacteria?
3. Write a report on Joseph Lister or Louis Pasteur. How did their scientific investigations relate to bacteria?
4. Who was Typhoid Mary? Did she actually exist?
5. What is Botulism? Why is it necessary to take special precautions to eliminate it?

6. Explain the process of home canning. How may home canning create a danger?
7. Hay bacillus gives off tremendous amounts of heat. Set up a hay culture. Mix hay and water in a beaker. Let it stand for several days. Take daily temperatures. Graph the results.
8. Set up an experiment to see which product spoils the quickest—frozen meat or fresh meat.
9. Does bacteria decrease with soil depth? Take soil samples at different levels to find out.
10. Which of the following environments do bacteria prefer: decaying leaves, freshly cut grass, gutter moisture, or vacuum cleaner dust?

chapter eight

10 ACTIVITIES THAT TEACH ABOUT PARASITES AND SAPROPHYTES

OVERVIEW

Students examine the world of parasites, hosts, saprophytes, and organisms who live together successfully. Investigations range from studying live specimens to extracting the intestinal tract of a termite.

Pupils dissect and examine roundworms, flukeworms, frogs, mosquitoes and flies. They scour the school campus for evidence of parasitic plant invasion. Students collect and investigate gall and scale damage.

Use the blackboard or overlay projector to diagram the meaning of these terms: Parasite, pathogen, host, carrier, symbiosis, commensalism, saprophyte, and mutualism. Here's a suggestion: Sketch a large letter P and a large letter H. P stands for parasite; H for host. Make cartoon drawings illustrating the 8 listed terms.

Stress the following points:

* A parasite may live on or in another organism at that organism's expense. Example: A tapeworm feeding in the host's intestine.
* Parasites which cause the host considerable damage are known as pathogens. Example: A lamprey sucking body juices from a fish.
* A host is the organism which houses the parasite. Example: a dog carrying fleas.
* A carrier is a host which passes parasites to other hosts. Example: A fly eating food, leaving bacterial parasites behind.
* Symbiosis is an interrelationship between 2 organisms of different species. Example: The lichen plant—A combination of both algae and fungus.

* Commensalism is a relationship in which the host remains unharmed and its visitor benefits. Example: The small fish, remora (visitor), attaches itself to the shark (host). The remora eats remaining food scraps.
* Mutualism occurs when 2 or more organisms living together benefit. Example: Flagellates living in the termites stomach break down cellulose into digestable food particles.
* Saprophytes are plants which live on waste products or non-living materials. Example: Lichen plant.

ACTIVITY 119

Are Some Anthropods Parasitic? A Student Experiment

Part A.

As mentioned earlier, a centipede will turn cannibalistic if the food supply diminishes. Do centipedes parasitize one another? Have students place 3 or 4 centipedes in a container free of food. Add moist shredded pieces of paper. Cover the container. Make small holes in the lid to allow sufficient ventilation. Tell them to check the containers each day for several days for evidence of parasitism.

ACTIVITY 120

Part B.

Have pupils place millipedes, centipedes, earwigs, and earthworms in a container devoid of food. Add moist shredded paper. Observe every day for evidence of parasitism. Give them these questions to answer:

1. Which animals demonstrate aggressive behavior? Answers will vary.
2. Which animals demonstrate non-aggressive behavior? Answers will vary.
3. Which animals seem to be the strongest? The weakest? Answers will vary.
4. Millipedes are generally considered plant eaters. Do they show carniverous behavior? Answers will vary.
5. Which animals, if any, show symbiotic behavior? Answers will vary.

Relate how some parasites live free in nature until a host becomes available. The parasites must eat enough to live, but not to the point of killing their host. In many instances, their survival depends upon the host's strength.

ACTIVITY 121

Dissecting An Ascaris

How do parasites travel from one host to another? Some develop resistant spore stages, wait for an unsuspecting host to eat them, and reproduce new organisms inside the host's body. Other stay active by flying or crawling to new locations.

Emphasize parasite mobility by showing students diagrams or slides of different parasitic life cycles, e.g., hookworms, tapeworms, trichina worms, etc.

Introduce nematodes, the free-living or parasitic multicellular animals which live in animals and plants. Let pupils investigate Ascaris, a parasitic roundworm which lives in pig and human intestines.

Pass out preserved specimens of Ascaris (the preserved specimens referred to in this unit can be ordered from Ward's Scientific Supply House). Be sure pupils understand not all roundworms are parasitic. Some are harmless; others create mild or serious disturbances. Have them place the Ascaris in a pan and observe carefully. Assign the following questions:

1. What is the shape of Ascaris? Long, thread-like body.
2. What kind of body structure does it have? Body has smooth, tough outer covering.
3. Where do you think the roundworm gets its name? From its cylindrical shape.
4. Are roundworms segmented? No.
5. Examine the Ascaris under a dissecting microscope or hand lens. How would you describe the head? Three rounded lips surround the mouth.
6. Cut the Ascaris lengthwise. What does the digestive tract look like? The digestive tract is a straight tube with the mouth and anus at opposite ends.
7. Where can Ascaris be found? Are they harmful? They can be found in human and pig intestines. Some may cause serious illness or death of the host.

ACTIVITY 122

A careful examination of moist soil may reveal nematodes resembling wiggly fine threads. Take students on a field trip around the campus. Have them collect samples of damp soil, wilting leaves, and plant roots. Let them inspect these items under a microscope or hand lens for nematodes.

Demonstrations And Experiments

1. Demonstrate total destruction by a "parasite." Place a small piece of meat in a test tube half full of dilute hydrochloric acid. Tell students the acid represents the parasite; the meat acts as host. Ask pupils to describe what happens to the parasite when the host dies.
2. Ask pupils to list examples of parasitic damage found around the school. Tell them to determine extent of parasitic damage. (This is difficult because the extent of host damage isn't easy to determine.) Look for leaf diseases.
3. How are certain parasites an important link in food chains? Have students make diagrams or sketches showing these relationships.
4. Give pupils preserved lamprey or leech specimens to examine. Tell them to inspect the head and mouth region. How are these structures adapted for parasitism?
5. Have students remove the nephridia (excretory organs) of a fresh squid. Give them textbook illustrations to guide their dissection. Tell them to make microscopic slides from the tissue and look for parasitic Dicyema—thin, threadlike worms.

STUDYING THE GALL INSECT

Discuss the parasitic role of gall insects. Explain how certain plant or animal parasites cause plant tissue to swell. Gall insects, e.g., flies, wasps, beetles, etc., produce various shaped galls. Some are round and conical (Figure 47); others flat and square. The female insect travels to a plant, lays her eggs on a leaf or branch. The larva develops while devouring the surrounding plant tissue. This disturbance creates abnormal tissue growth and division of plant cells. These conditions provide a suitable home for the gall insect (Figure 48).

ACTIVITY 123

Go on a gall hunt. Take the class around the campus. Inspect trees, especially oak and cedar. Check the crowns of trees and shrubs.

Part A.

Collect as many varieties as possible. Include old discarded galls. Bring materials into the classroom.

Figure 47

Figure 48

Part B.

Cut open each gall swelling (Figure 49). Examine the interior with a hand lens or microscope. Describe, including a sketch, what you find.

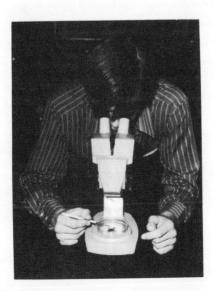

Figure 49

Have students answer the following questions:

1. How many different shapes are noticeable? Answers will vary.
2. How do old and new galls differ? Old galls are dried out and usually contain a hole where the insect left the nest. New galls possess living plant tissue surrounding insect larvae.
3. Does more than one insect live in the same nest? Answers will vary.
4. How many gall insects did you find? Answers will vary.
5. How are galls injurious to the plant? Gall insects drain the juices from the plant.

INSPECTING THE SCALE INSECT

Introduce students to Family Coccidae, the scale insect. If possible, show picture illustrating examples of scale invasion (Figure 50). Mention the following points:

1. Scale insects, Order Homoptera, are small to minute.

Figure 50

2. The adult male has no mouth and cannot feed. He lives just long enough to fertilize the female's eggs.
3. The female does the damage. The adult female has no eyes, legs, or wings. She has a rounded shape, and is covered by secreted wax or shell-like scale.
4. The female stays in one spot for most of her life.
5. The scale causes great damage by sucking juice through tiny tube-like beaks.

ACTIVITY 124

Is the school or community inhabited by scale? Find out. Let students examine the school grounds, the block where they live, and sections of the community for evidence of scale invasion. Have them bring specimens into the classroom and examine under a hand lens or microscope. Pupils who collect scale during the spring may find numerous eggs attached to the scale's abdomen. They appear as a fine, white powder.

Place several collected specimens in a small bottle with a screw cap (any small covered container will suffice). Mail to the State Department of Agriculture. Provide date and location where insects are found. Ask Department for a report, e.g., what specie of insect, how is it destructive, methods for extermination, etc.

EXAMINING THE HEADS OF MOSQUITOES AND HOUSE FLIES

Pupils know the nasty things flies and mosquitoes do. However, a review of their careless feeding habits helps explain the way they spread disease. Mention how mosquitoes transmit yellow fever, sleeping sickness, filariasis and malaria by piercing human skin with their tubular mouths. They pierce the skin, suck the blood, and pass parasitic organisms into the host. The house fly's sponging mouth parts may contaminate food and spread disease.

ACTIVITY 125

Pass out freshly killed mosquitoes and house flies. Have students remove the head from each specimen and examine mouth parts under a hand lens or microscope. Tell pupils to sketch each mouth part and study the structural differences.

Suggested Problems And Questions

Assign pupils these questions:

1. Collect fleas from a dog or cat. Bring them into the classroom. Examine under a microscope. How are their mouth parts adapted for sucking blood from a host?
2. Why may it be unsafe to eat raw beef or pork? What parasite might be present?
3. Examine an oak tree for mistletoe. Mistletoe is considered parasitic. What percent of the tree is covered with mistletoe? How may mistletoe be injurious to the host tree? Do you think the mistletoe will eventually kill its host?
4. Make a list of parasites who are themselves parasitized. For example, fleas carry intestinal parasites.
5. List 5 true parasites which use man as a host.
6. Diagram the pork tapeworm cycle. List several ways the tapeworm can be stopped from developing in a pig.
7. Athlete's Foot, a fungal parasite which grows on the skin of feet, is extremely difficult to kill. Why is this so?
8. Examine the gills of a freshly killed fish. Are parasites present? Check other parts of the fish—heart, intestines, and muscles—for parasitic invasion. Are these organisms true parasites?
9. Many insects carry parasites. Examine the intestines of millipedes, centipedes, ground beetles, and earwigs for parasitic organisms.
10. Examine the pincers of a freshly caught crayfish for white, worm-like organisms. Would you classify these organisms as parasites? Why or why not?

DO FROGS CARRY PARASITES? A STUDENT EXPERIMENT

Mention how many parasites thrive in warm moist areas. Some are external, living on the outside of the host; others seek internal rewards within the host. Few creatures escape the ravages of hungry parasites. Amphibians are no exception.

ACTIVITY 126

Provide pupils with freshly killed frogs. Have them perform the following exercises:

Part A.

Dissect the liver and lungs. Tease the tissue apart. Look for parasitic worms. Transfer tissue (and parasites) to a Petri dish containing saline (Ringer's) solution. Examine organisms under a microscope.

Part B.

Remove the intestinal tract. Cut it open lengthwise. Tease tissue away from the inside wall. Transfer tissue to a microscopic slide. Add a drop of saline solution to the tissue. Have students sketch any organisms they find.

Part C.

Scrape tissue from around the rectal opening. Repeat the procedure in Part B. Look for protozoan organisms. Have pupils sketch these organisms.

Assign the following questions:

1. Do all the protozoan organisms look alike? If not, how do they differ? Probably not. They may differ in size and shape.
2. Are all these organisms parasitic? Explain your answer. This is very difficult to answer from miscroscopic examination.
3. Where can the greatest number of organisms be found in the frog's body? Answers will vary.

LOOKING INSIDE THE FLUKEWORM

Use diagrams to explain the life cycle of the liver fluke. Parasitic life cycles interest students from the standpoint of physical development, host contact, distribution, and inflicted damage.

Emphasize these main ideas:

1. Flukes require a different host for each stage of development.
2. Flukes travel to specific organs. They attack the liver, blood, and intestines.
3. A typical example, the Oriental liver fluke, runs a complex life cycle. Fluke eggs leave the liver tissue of man, pass through feces and become ingested by a snail. The larvae matures within the snail's body, develops a tail, leaves the snail, and enters a fish. The larvae beds into the flesh of the fish. A person who eats raw or poorly cooked fish may open the door to new invasion.

ACTIVITY 127

Pass out preserved specimens of flukeworms. Have pupils carefully examine the external and internal structures. Give no specific directions. Allow students to make their own observations.

Assign these questions:

1. Is the outside skin thick or thin? A thick cuticle covers the body.
2. Does the fluke possess cilia (hairlike projections) over its body? No.
3. How does the fluke's internal structures compare with that of Ascaris? Ascaris has a less complicated internal structure.
4. What structures allow the fluke to attach itself to a host? A sucker type or hook device.
5. A fluke may lay thousands of eggs each day. Why is this necessary? For survival purposes.
6. The liver fluke needs a water snail to carry on its life cycle. What may happen if a snail host isn't present? The fluke larva may cease to develop.
7. How many hosts are necessary to carry on Oriental liver fluke development? Three. Man, water snail, and fish.

Emphasize many organisms live comfortably together. Each, in its own way, adds to the subsistence of the other. Such a relationship is known as mutualism. Here are some excellent examples:

1. A certain species of ant carries aphids to their tunnels to eat grass roots. In return, the aphids secrete a sweet liquid for ants to consume.
2. The lichen, a combination of alga and fungus, grow together harmoniously. The green algae provides the food; the fungus anchors the plant.
3. The bee lands on a flower, collects nectar, and leaves pollen behind to aid plant production.

ACTIVITY 128

What's Inside A Termite's Intestine? A Student Experiment

Mention how termites depend upon certain species of protozoa (flagellates) to digest the cellulose they eat. Since termites cannot break down cellulose, the protozoa living in the termite's intestine does the job for them. The protozoa eats the wood, changes it to soluble carbohydrates which termites use.

Provide students with freshly killed termites. Have them carefully remove the intestinal tract on a slide containing a drop of saline solution. Place a cover slip over the slide. Add a drop of methylene blue dye. Observe organisms under high power.

Offer these questions:

1. How would you describe the organism's shape?
2. How do you think these flagellates move about? By the tail or whip-like structures surrounding the body.
3. Would you say the flagellates are slow or fast moving? Answers will vary. However, they are usually fast moving.
4. Why is the termite - flagellate relationship considered an example of mutualism? The flagellate provides food for itself and the termite.
5. Do you think wood-eating termites can live without flagellates? Not for very long.

BRINGING SAPROPHYTES INTO THE CLASSROOM

Students are familiar with the Hollywood version of creatures existing off the dead bodies of other organisms. These ghoulish tales keep the adrenalin flowing. A plant which thrives on waste material or the dead bodies of other organisms is called a saprophyte.

ACTIVITY 129

Allow students to collect saprophytic samples under rocks, logs and compost piles containing moldy leaves. Have them examine materials with a hand lens or microscope.

TEN SUGGESTIONS FOR FURTHER STUDY

Give interested students the following suggested problems:

1. Collect discarded crab shells. Check the back, legs and pincers for barnacles. What percent of the examined structure is covered by barnacles? Which part of the crab seems to carry more barnacles? Find out if crabs carry parasitic barnacles.
2. The human or "beef tapeworm" rests, eats and reproduces within the intestines. Research the following items:
   a. How do proglottids aid in tapeworm growth and reproduction?
   b. How does the scolex benefit the tapeworm?
   c. What countries show a large percentage of tapeworm invasion? How do you account for this?
   d. Explain how tapeworms can be transmitted to humans via dogs or fish.
3. Make a display of saprophytes. Grow citrus fruit and bread mold. See how many different varieties you can raise.

4. Give examples of predators and their victims. How does this differ from parasitism? What role do predatory animals play in food chains?

5. Describe the damage done by the Japanese beetle. Why is this insect so dangerous?

6. When two organisms derive some benefit from each other, it is called symbiosis. Give examples and tell how the symbiotic relationship occurs.

7. Some organisms are both free-living and parasitic. List examples and describe the stage of life, i.e., larvae or adult, these organisms are parasitic.

8. List the various ways parasites enter their hosts.

9. Locate aphids feedings on a rose bush. Make daily observations. Assess the damage to the host plant.

10. Describe how mites and ticks parasitize warm-blooded hosts.

chapter nine

29 ACTIVITIES THAT EXPLAIN IMPORTANT FACTS ABOUT THE ANIMAL KINGDOM

OVERVIEW

The animal kingdom creates much excitement. Students first encounter members from the invertebrate world. They study the clam, squid, earthworm, crayfish and insect.

Two members of the vertebrate group, fish and frog, visit the classroom. Pupils examine external body parts, dissect internal structures, and observe the behavior of living specimens.

The clam, a "soft-bodied" animal, belongs to the phylum Mollusca. Mollusca, the second largest invertebrate group, appears on the menu of many fine restaurants. Other members include snails, slugs, oysters, squids and octopuses.

ACTIVITY 130

Dissecting A Clam

Pass out fresh or preserved clams. Either salt water cr fresh water specimens work well. Provide students with dissecting materials, trays, hand lens or microscopes, microscopic slides, slips, sketch paper, reference texts and diagrams.

Part A. External Features

Give pupils these guidelines:

1. Place a clam in a tray. Carefully examine the outside shell.
2. Sketch and label: Umbo, lines of growth, periostracum, dorsal, ventral, posterior and interior areas.
3. Peel away strips of periostracum. Examine under a microscope. Describe what you see.

Part B. Internal Structures

1. Pry the clam shell apart.
2. Locate, examine and sketch these structures: Incurrent siphon, excurrent siphon, anterior adductor muscle, posterior adductor muscle, mantle, and foot.

Part C.

Have students remove the stomach and intestines. Advise pupils to thoroughly inspect stomach contents. Tell them to make several microscopic slides of stomach and intestinal scrapings.

Assign students the following questions:

1. What do lines of growth indicate? Intervals between successive growth stages.
2. How does a clam use its foot? As an anchor to draw its body forward.
3. What is the purpose of the excurrent siphon and incurrent siphon? Water comes through the incurrent opening, passes by the mantle cavity, and leaves through the excurrent opening. Food particles enter the clam through the entering water.
4. What is the purpose of the periostracum? Gives protection to the calcareous shell from being dissolved by carbonic acid in the water.
5. What is the function of the posterior adductor muscle and anterior adductor muscle? They draw the valves together.
6. What do you think is the oldest structure on the clam? Umbo. The animal grows away from the umbo.
7. What type of food do clams eat? Mostly microorganisms and minute organic particles.

DISSECTING A SQUID

ACTIVITY 131

Students enjoy hearing tales which pit the ferocious sperm whale against the sixty foot giant squid. These raging battle stories inspire pupils to carefully inspect the tentacles, jaws, suckers, and ink sac of the squid.

The comical-looking squid, whose "arms" or tentacles are wrapped around the head like the mythical Gorgon, belongs to the class Cephalopoda which means "head-footed."

Part A. External Features

Tell pupils to:

1. Place a preserved or fresh squid in a tray.

2.   Remove an arm of the squid. Count the number of suckers.
3.   Remove a sucker. Place under a microscope or hand lens. Sketch and label: Muscular stalk (point of attachment) and the toothed, horny ring lining.
4.   Run your fingers along the squid's body. Poke at intervals. Describe the body texture.

Part B. Internal Structure

1.   Remove the jaws and radula. These structures can be found slightly forward of the eyes in the mouth region. (Refer to squid diagram showing internal structures.)
2.   Notice how the jaws resemble parrot beaks. Feel them. The radula, rows of very fine teeth, lies behind the jaws. Study them under a microscope. Make a sketch of your observation.
3.   Carefully remove the ink sac which lies near the anus. Extract the ink with an eye dropper. Transfer the ink to a beaker of clear water, one drop at a time. Describe how each drop spreads through the water.
4.   Remove the feathery-shaped shell which lies under the mantle of the anterior surface. Closely examine the shell. Twist or bend it. List several outstanding features of the shell.

Give pupils the following questions:

1.   What is the function of the radula? Draws food into the mouth and tears it apart.
2.   What is the purpose of the ink sac? It acts as a "smoke screen" or diversion tactic.
3.   How does the cellophane-like shell aid the squid? This light, flexible shell allows the squid to make fast, quick-turning movements.
4.   What structures help the squid move rapidly through the water? Water enters the mantle cavity and leaves through the funnel opening (located in head region). When water is quickly forced out of the funnel, the squid moves rapidly through the water. The fins help guide the squid.

CUTTING OPEN THE EARTHWORM

Few students miss the opportunity to thread a slimy worm on a fish hook or practice broken-field-running around worms after a rain.

Practically everyone is familiar with the phylum Annelida which includes earthworms, sandworms and leeches.

ACTIVITY 132

Provide students with preserved earthworms (Lumbricus terrestris), dissecting materials, trays, microscopes or hand lens.
Have pupils do the following:

1. Lay an earthworm in a tray.
2. Grab the head end with one hand, the tail end with the other. Straighten out the worm. Pin down both ends.
3. Sketch your specimen.
4. Remove the pins. Pick up the specimen and examine it.

Answer these questions:

How does the ventral (belly) side differ from the dorsal (back) surface? Answers will vary.

How many somites or body segments does your specimen have? Answers will vary.

Describe the structure between somites 31 and 37. The clitelum, or pack saddle, is a smooth glandular swelling. It resembles a collar.

Run your index finger and thumb down the sides of the worm. How does it feel? Rough and prickly.

How can you tell the posterior end from the anterior end of the worm? A fleshy lobe overhangs the mouth. The clitelum is closer to the head end than the tail end.

ACTIVITY 133

Part A.

Let students perform an "autopsy." Provide diagrams or text references which illustrate internal structures.
Have pupils do the following:

1. Divide a paper towel into 5 parts.
2. Remove the structures listed in step 4 and lay each in a different section of the paper.
3. Label each section.
4. Structures:
   "Brain" (between somites II - IV)
   Pharynx (between somites IV - V)
   Crop (between somites XV - XVI)
   Gizzard (between somites XVII - XVIII)
   Intestine (between somites XXI - Anus)

Part B.

Have pupils examine each structure under a hand lens or dissecting microscope. Assign these questions:

1. What color is the "brain" tissue? Grayish-white.
2. What is the function of the pharynx? A passageway which connects the mouth with the esophagus.
3. What is the function of the crop? (Cut the crop open. Examine the contents.) The crop is a food storage compartment. Food particles may be present.
4. Cut open the gizzard. Describe what you find. What is the function of the gizzard? The gizzard is a muscular structure found in the intestinal tract. It helps the worm grind swallowed food particles. Students may find small stone pieces inside the gizzard* which aid the worm in grinding food.
5. What happens to food in the earthworm's intestine? Food is digested and absorbed.

WAYS TO STUDY A LIVE EARTHWORM

Live earthworms lend themselves to fascinating study. Chapter One lists several exercises which stress observing earthworm behavior. The next 3 activities offer further investigations on the living earthworm.

ACTIVITY 134

Provide pupils with several large glass jars or beakers. Divide them into small groups. Give these guidelines:

1. Fill 2 containers with moist soil or coffee grounds.
2. Place several worms (6 - 12) in each container.
3. Set one beaker under a strong light. Place the other in a cold area, preferably in a refrigerator.
4. Wait 20 minutes. Then place the containers together on the laboratory table.
5. What behavioral differences do you observe between each group of worms? Answers will vary. Generally speaking, long-lasting cold temperatures drive them deep into their burrows. Warm temperatures have the same effect.

---

* The gizzard of a chicken serves the same purpose. If possible, cut open and examine the contents of a chicken gizzard.

ACTIVITY 135

Give each student group a shoe box or tray. Have them separate the container into 4 equal sections with cardboard inserts. Have them place these soils in different sections:

    a. Moist, sandy soil
    b. Moist leaves and grass
    c. Dry, sandy soil
    d. Dry coffee grounds

Drop 10 worms into the box. Set the box in a cool, dark area of the room. Check the next day to see which material most worms prefer.

ACTIVITY 136

Repeat the procedure for Activity 135. Use the same worms. Replace soil with moist topsoil. Add the following ingredients to the different sections:

    a. Pieces of celery or celery salt
    b. Thin shreds of cabbage
    c. Carrot leaves
    d. Grass clippings

Set the box in a cool, dark area of the room. The next day find out what food material the earthworms prefer.

Give students these questions:

1. What do you think determines food preference in earthworms? The possibility of taste cells in the mouth and pharynx.
2. What do you think determines soil preference in earthworms? Worms possess light-sensitive cells at the anterior and posterior ends of its body. Scientists believe earthworms use "touch" cells, which occur all over the body, to pick up temperature changes and chemical stimuli.
3. Will only one experiment provide conclusive results regarding animal behavior? No. Many repetitions are generally more reliable.
4. What benefits can be gained from doing these experiments? An opportunity to observe behavior which occurs at that particular moment.

METHODS OF STUDYING THE CRAYFISH

Many crawling creatures with segmented bodies and jointed appendages wear the phylum Arthropoda name tag. Horror film producers spend considerable money, time, and energy enlarging the external features of crabs, insects, spiders, centipedes, and scorpions.

These cumbersome critters make classroom investigations difficult. So, for practical purposes, let your students begin their study with a normal sized crayfish.

ACTIVITY 137

Pass out preserved crayfish. Provide students with dissecting materials, reference books and diagrams.

Part A.

Lay a crayfish in a dissecting pan. Make a sketch and label: Eye, antennule, antenna, cheliped, walking legs, cervical groove, telson, abdomen, uropod and carapace.

Part B.

Remove and dissect the stomach and intestines*. Describe what you find. Can you identify any food particles? Make several microscopic slides of stomach contents.

ACTIVITY 138

Let students discover how crayfish respond to various stimuli. Obtain several live crayfish. Keep them in a fish tank supplied with an air pump.

Part A.

Have pupils test the righting reflex (ability to turn right side up) of the crayfish. Provide these materials for each student group: 2 small crayfish, a cardboard or plywood strip, approximately 11 1/2 by 8 inches.

Offer these guidelines:

1. Mark each crayfish with felt pen or tape. Make sure you can identify each one.

---

* If possible, let students examine the internal structures of freshly-killed crayfish. Food particles are easier to recognize.

2.   Place the first crayfish, Subject A, in a tray on the laboratory table. *Caution:* Beware of the flesh-seeking pincers. Be sure the dorsal side of the carapace touches the tray bottom. Record which side of the body the crayfish rights itself and turning time. Carefully observe which structures the crayfish use to turn over. Make 6 trials of 2 minutes duration with 1 minute rest periods between trials.
3.   Switch subjects. Repeat procedure.
4.   Calculate the average righting time and preferred righting side for each subject.

Part B.

1.   Repeat Part A procedure. This time hold the cardboard or plywood slab at a 20 degree angle. Place each subject, one at a time, with its head pointing away from the apex.
2.   Record righting time and preferred turning side.
3.   Repeat several trials.

Discuss these observations with the class. Bring up the following points:

*   These experiments prove nothing. What side a crayfish prefers in righting itself may be determined by age, health, strength, fatigue, laboratory surroundings, or a combination of these factors.
*   Statocysts, balancing organs, may or may not be affected by turning the crayfish on its back or pointing its head away from or toward the apex.
*   The large pincers, swimmerets, walking legs, and abdomen all work together in righting the crayfish.

ACTIVITY 139

Do crayfish show habituation to stimuli? Have students choose 2 subjects and try the following experiment:

1.   Place the crayfish, one at a time, on a tray on the laboratory table.
2.   Slowly move a pencil, a piece of glass tubing, etc., toward the subject's head. Stop when you come within 5 millimeters of its head.
3.   Repeat presenting and retracting the stimulus between 40 and 50 times.
4.   Try other stimuli, e.g., pencil flash light, colored crayons, and so on.

Emphasize how difficult it is to measure certain behavior. The crayfish may stop reacting because of fatigue or habituation. There's no way to tell for sure.

ACTIVITY 140

Have pupils place several (4 or 5) crayfish together in a large tank of water. Let them observe behavior. Here are some guidelines:

1. Do crayfish prefer a certain area in the tank?
2. Do you see signs of aggressive behavior? If so, describe the behavior.
3. How do crayfish demonstrate dominant behavior? Passive behavior?
4. Vigorously stir the water with a dip net or glass stirring rod. Does turbulent water have any effect on crayfish behavior? Explain your answer.

Suggested Problems And Questions

Offer students the following items:

1. Collect some garden snails. Sprinkle talcum powder or chalk dust on the bottom of a glass tray. What kind of trail does the snail leave? Do all snails make a similar trail? What factors do you think determine snail trails?
2. Let garden snails crawl over different surfaces, e.g., smooth, bumpy (sand paper), pitted, or wavy. How do snails adjust to these surfaces?
3. Place empty clam shells in 2%, 6%, 10%, and 20% carbonic acid solutions. What effect do these solutions have on clam shells? What happens to the periostracum (outside protective covering)? Continue to soak clam shells. How long must they soak before the periostracum dissolves? Increase the carbonic acid solution. What happens?
4. List several ways earthworms benefit farmers. How can earthworms become destructive?
5. Why do earthworms come to the surface after a hard rainfall?
6. Earthworms do not have eyes. How are they able to move about?
7. The circulatory and nervous system of an earthworm is considered primitive. Why is this so?
8. Study and compare different preserved arthropods, e.g., grasshoppers, spiders, millipedes, and centipedes. List the important features for each animal.
9. Take students on a campus field trip. Have them collect several

spiders and return to the classroom. Tell them to transfer spiders to a Petri dish, cover it, and place under a microscope. Assign these questions:

How many distinct body parts does the spider have?

How many walking legs do they have?

What do you see at the tips of each walking leg? What is their purpose?

What structures surround the hard outside body covering?

Sketch a walking leg. Label the coxa, trochanter, femur, patella, tibia, metatarsus, and tarsus.

Refer to a textbook. Sketch a human leg. Label each of the foregoing structures.

Have pupils release the spiders. Remind them spiders benefit man by feeding upon harmful insects.

10. Where are the statocysts (balancing organs) located in the clam, squid, and crayfish?

11. What members of phylum Mollusca, phylum Annelida, and phylum Arthropoda do people enjoy eating? How are these morsels prepared?

Electrical experiments stimulate students and animal subjects as well. Are crayfish affected by electricity or light? Does their hard carapace protect them from shock and heat? The next activity will help provide the answers.

Provide students with glass trays, electrical power units, copper electrodes, and live crayfish. Divide students into small groups. Check with electrical shop teacher for available power units (1 to 10 volt regulator) and methods for preparing tray water.

ACTIVITY 141

Have pupils do the following:

1. Place 1 or 2 crayfish in a glass tray.
2. Add water to the tray. Pour enough to cover the animal's body.
3. Extend electrodes into the water. *Caution:* Warn students not to play with electrodes after they've been in water. Have them unplug electrical power units when they finish the experiment.
4. Gradually induce shock, one volt at a time, until the crayfish responds to the electricity.
5. Set up a chart and list the type of behavior elicited for each volt of electricity.

ACTIVITY 142

Provide student groups with 75-watt light bulbs, light sources, trays, and fresh crayfish. Give these guidelines:

1. Place a crayfish in an empty tray.
2. Provide cover, e.g., dark paper, a book, etc., over one end of the tray.
3. Set up the light apparatus to shine directly over the crayfish. Keep the bulb approximately 3 inches above the crayfish.
4. Record 10 trials, 2 minutes each, with 1 minute rest periods.
5. Set up a chart and list the type of behavior elicited for each trial.
6. Repeat procedure with crayfish submerged in water. Does the crayfish react in the same way?

Several million species of insects roam the earth. Some bite, sting, and suck man's blood. Others eat his crops, harass his animals, and infect him with disease. However, aesthetically speaking, insects add to man's pleasure in life. Crickets chirp a summer song, butterflies spread their beautiful wings, and glowworms brighten a summer night. And insects form an important link in the food chain.

DISARTICULATING THE GRASSHOPPER

ACTIVITY 143

Provide students with diagrams or textbooks illustrating insect anatomy, dissecting materials, trays, preserved grasshoppers, hand lens or microscopes.

Give the following directions:

Part A.

1. Place a preserved grasshopper in a tray.
2. Carefully inspect the head, thorax, and abdomen.
3. List and describe the structures associated with each body division.

Part B.

1. Cut off the head of the grasshopper. Carefully remove a compound eye, antenna, upper lip (labrum), jaws (mandibles), accessory jaws (maxillas), and lower lip (labium). Examine each structure under a hand lens or microscope.
2. Make a sketch and give the function for each structure.

Part C.

1. Remove both pairs of wings. Closely examine each wing. Describe the texture of the first, or outside wing. Is it suitable for flying? What is its purpose?
2. Examine the second, or inside wing. Describe its texture. How does the grasshopper use this wing?
3. Weigh each wing. Which one is heavier? By what per cent? How do you account for this difference?
4. Unfold the inside wing. Trace its outline on a piece of paper. What airplane structure has this shape?

ACTIVITY 144

Have pupils do the following:

1. Lay the thorax and abdomen of the grasshopper in a tray. Remove the legs.
2. Locate, inspect, and sketch these leg parts: Coxa, trochanter, femur, tibia and tarsus.
3. Carefully inspect the chitinous shield which covers the thorax. What major structures attach to the thorax? What is their function?
4. Remove the food gut (esophagus, crop, gizzard, stomach, intestines, rectum and anus).
5. Cut the crop open. Examine the inner walls. What purpose do they serve? What is the function of the crop?
6. Inspect the inside of the stomach. Can you identify any food particles?
7. Remove the gizzard. Examine the gizzard lining. What purpose does it serve?
8. Remove the intestine. Stretch it out in a tray. Measure its length. How does this measurement compare with the grasshopper's total length (head to rear)?

Pass out sheets describing the different orders of insects. Go over the major divisions with the students. Use available insect collections, illustrations, slides, or filmstrips to supplement your discussion.

COLLECTING INSECTS AND PREPARING COLLECTIONS

ACTIVITY 145

Let students spend one or two classroom periods making their own insect collecting materials. (Figure 51) Many excellent science texts offer simple directions for making equipment (see *Exploring Life Science*, Thurber and Kilburn, Allyn and Bacon, Inc., 1966, pp. 50-54).

Figure 51

ACTIVITY 146

Discuss and demonstrate how to collect and pin insects. Pass out insect pins, scissors, and index cards. Here's a suggestion: Set up the overhead projector. Sketch various insect shapes on a transparency sheet. Show the proper method for pinning different insects.

Have students cut out insect body shapes, e.g., beetles, bugs, flies, butterflies, grasshoppers, and so on. (Figure 52.) Tell them to pin each cut out properly. Also, let them practice making labels and pinning small insects.

ACTIVITY 147

Have pupils spend the next few days collecting insects around the campus or home. Divide the time between collecting, preparing, and identifying insects. Encourage students to keep a collecting log (Figure 53). This log gives the insect number, date, location where collected, and what the insect was doing when collected.

Suggested Problems And Questions

Offer these items to interested students:

1. Compare a beetle or bug with a spider. How does a spider differ from an insect?

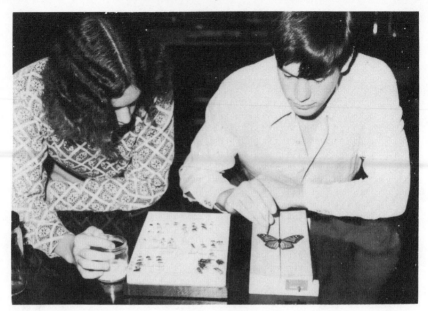

Figure 52

Figure 53

2. Collect different stages of ladybird beetle larva. Describe each stage, including growth development, color change, movement, etc.
3. Go on an egg hunt. Collect insect eggs. Let them hatch.

4. Which insects practice protective coloration? Search for insects which blend in with their environment.

5. How are water striders able to walk across water? Observe the legs of a preserved specimen under a hand lens. How are their legs adapted for this purpose?

6. Examine moth and butterfly antennae. How are they different? How are they alike?

7. Make a microscopic slide of butterfly wing scales. Describe what you see.

8. An insect changes form from one stage of development to another. This is known as metamorphosis. Describe gradual, incomplete and complete metamorphosis. List several insects which undergo these changes.

9. How do the following insects have mouthparts adapted for their environment?

Grasshoppers	Butterflies
Lice	Beetles
Aphids	Bees

10. Ants and bees are social insects. They form a caste system. What does this mean?

11. Some insects have unique features. List distinct features for each of the following insect orders:

Protura	Hemiptera (true bugs)
Collembola	Diptera (true flies)
(Springtails)	
Dermaptera	Coleoptera (beetles, weevils)
(Earwigs)	
Odonata	Hymenoptera (ants, wasps, bees)
(Dragonflies)	

12. Halters, short knobby appendages, replace the hind wings in dipterous (order Diptera) insects. They extend from each side of the thorax and act as balancers. Remove the halters from a crane fly, house fly, or mosquito. What effect does this have on insect flight?

INVESTIGATING THE STRUCTURES OF A FISH

Begin with a discussion of the unique characteristics of a fish. Mention these points:

1. Fish swim in water which ranges from cold fresh water to warm salt water.

2. They breathe by means of gills.
3. They possess fins which differ in size and shape. Fins enable a fish to change direction, attain balance, speed up or slow down.
4. They are sensitive to vibrations, have a well-developed sense of taste, but do not see very well.
5. Many have bony skeletons (except sharks, rays, and sturgeon) with rough, smooth, and heavy scales.

ACTIVITY 148

Pass out preserved fish. Yellow perch make excellent specimens. Have pupils place a fish in a tray, then do the following:

1. Locate, inspect, sketch, and label these structures:

mouth	lateral line
eyes	pelvic fin
operculum	anus
pectoral fin	anal fin
dorsal fins	caudal fin

Give students the following questions:

1. How would you describe the body shape? Long and narrow. Higher than wide.
2. Where are the nostrils located? On the head between the upper lip and eyes.
3. Does the fish have eyelids? No.
4. How do the two separate dorsal fins differ? The first dorsal fin (near head) has rigid spines. The second dorsal fin (near tail) is soft.
5. Lift the operculum. What lies beneath this structure? Gill structures.
6. How many spines does the first dorsal fin have? Between 13 and 15.
7. Pull and twist each fin. Which one is rigid? The first dorsal fin.
8. Trace the lateral line. How long is it? Extends from the tail to the top of the operculum.
9. Inspect the mouth. What structures are present? Teeth and tongue.
10. Rinse off your fish. Wipe it dry. Place the fish in your hand. Guess its weight. Now weigh the fish. How close was your estimate?

ACTIVITY 149

Have students carefully remove the head. Caution them not to destroy gill structures. Tell pupils to remove the brain and lay it in a tray.

Give these guidelines:

1. Locate, inspect, sketch and label the following structures:

cerebellum	spinal nerves
optic lobes	medulla
cerebral hemispheres	spinal cord
olfactory lobes	

Assign pupils the following questions:

1. How does the cerebellum aid the fish? Helps fish coordinate muscular movements and maintain balance.
2. The olfactory lobes are located in the fish's snout. What is their purpose? Smell.
3. What is the function of the cerebrum? To help the fish learn.
4. What is the function of the optic lobes? Sight.
5. What are the nerve fibers located along the side of the brain called? Spinal nerves.
6. Weigh the brain. The brain makes up what percent of the fish's total weight? Answers will vary.

ACTIVITY 150

Have students remove an eyeball from the fish and place it in a tray.

Offer these directions:

1. Locate, inspect, sketch and label the following structures:

lens	optic nerve
iris	cornea

2. Remove the lens from the eyeball. Lay the lens over newsprint or a pencil mark. Does the lens magnify?

Give pupils these questions:

a. What is the function of the iris? Regulates the amount of light entering the eye.
b. What is the cornea? The outer transparent part of the eyeball.

c.   Where is the optic nerve located? What is its purpose? In the rear of the eye. It carries impulses of sight from the eye to the brain.
d.   How is the retina important to vision? It is made up of light-sensitive neurons.

ACTIVITY 151

Have pupils cut away the operculum, remove the gills, and place them in a tray. Tell students to sketch the gills and label: gill filaments, gill arch, and gill rakers.

Assign these questions:

1.   What is the purpose of the gill filaments? Gill filaments contain capillaries. Blood in the filaments passes off carbon dioxide and absorbs oxygen from the water.
2.   What is the purpose of the gill arch? To support the gills.
3.   What is the purpose of the gill rakers? To protect the gills and prevent food from passing out the gill slits.

ACTIVITY 152

Have students carefully remove the intestinal tract and all internal organs. These structures can be transferred to beakers containing saline solution.

Part A.

Tell pupils to place the air bladder (found between the stomach and kidney) in a large beaker of water.

Assign these questions:

1.   What causes the bladder to float? Gases.
2.   What gases fill the bladder? Carbon dioxide, nitrogen, and oxygen.
3.   How does the air bladder help the fish? It helps the fish adjust to different water depths.
4.   What effect does sudden temperature change have on fish? Sudden temperature change may force a fish's stomach out of its mouth.

Part B.

Have pupils remove the heart and lay it on a tray. Give these guidelines:

1. Locate, inspect, sketch and label the following structures:

         Ventral Aorta   Auricle

         Bulbus Arteriosus   Sinus Venosus

         Ventricle   Cardinal Vein

     Give students these questions:

1. Which structure receives blood from the body? Cardinal vein.
2. Where does the ventral aorta lead? To the gills.
3. How many chambers does the fish heart have? Two.

ACTIVITY 153

Let students examine the stomach and intestines for food particles. Have them sketch and label: Pharynx, esophagus, stomach, intestine and anus.

ACTIVITY 154

Discuss the importance of fish scales and how they differ. Use the scale chart as a guide (Figure 54).

Have pupils remove a perch scale and place it under a hand lens or microscope. Tell them to sketch the scale and describe its appearance.

Suggested Problems And Questions

Assign these items to interested pupils:

1. Read a fishing story from an outdoor magazine. How does fish behavior attract fishermen?
2. Compare scales from several different fish. How are they different? How are they alike?
3. Write a report on color changes in fish, i.e., how does a flounder adjust to its environment.
4. Why are some fish called "trash" or "rough" fish? Give examples.
5. Lampreys and hagfish are sometimes classified as fish. Is this a fair evaluation? Why or why not?
6. Annual rings formed on scales indicate the age of a fish. Examine different scales (carp scales are large and easy to examine). Try to determine age of fish.
7. How does the modern fish tail differ from its ancestor?

SCALE	CHARACTER-ISTICS	SKETCH	EXAMPLE
CTENOID	Rough with bumpy edges. Comblike margin.		PERCH
GANOID	Heavy, plate-like rows. Covered with enamel.	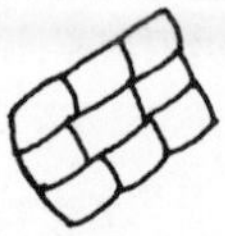	PRIMITIVE FISH
CYCLOID	Smooth tex-ture with concentric rings.	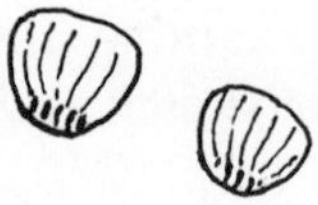	STURGEON
PLACOID	Hard, spiny structures; toothlike.	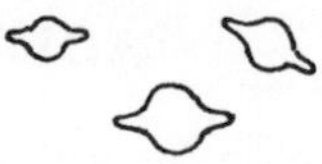	SHARKS

Figure 54

8. List several fish which can live in both fresh and salt water.
9. Shark attack reports occasionally appear in the newspaper. Collect these reports. List the conditions leading up to the attacks, e.g., what victims were doing, where they were attacked, what were the weather conditions, and so on.
10. Starfish and jellyfish are not true fishes. How do they differ from fish? Why does the term fish appear in their names?
11. How is the circulatory system, nervous system, and respiratory system advanced over invertebrates?

FIVE WAYS TO STUDY A FROG

The Class Amphibia includes toads, frogs, and salamanders. These clever animals are sandwiched somewhere between the fishes and reptiles. They live in water and roam on land. Their external and internal structures offer interesting study.

ACTIVITY 155

Provide dissecting materials, trays, hand lens or microscopes, and preserved frogs (leopard or bullfrogs).

Have students place a frog in a tray and make the following observations:

1. How does the ventral side differ from the dorsal side? The ventral side is smoother and lighter in color.
2. Feel the frog's skin. Describe its texture. The frog has smooth and moist skin.
3. Does the frog have scales? No.
4. How many walking or swimming limbs does the frog have? 4.
5. How many digits does the frog have on each limb? The forelimbs have 4 digits; the hind limbs have 5 digits.
6. What is unique about the hind toes? The hind toes are webbed.
7. Notice the large circular structures located behind each eye. What are they called? Eardrums or tympanic membranes.
8. How is the hind leg adapted for jumping? Well-developed muscles give the frog added strength.
9. How does the frog blend in with his environment? Skin coloration.
10. List other outstanding features present on the frog. Answers will vary.

Go over the answers in class.

ACTIVITY 156

Have pupils study the frog mouth. Tell them to cut the muscles at the corner of the mouth, pry it open by bending the lower jaw way back. They can use wooden splints to hold it open.

Give students these directions:

1. Locate, inspect, and sketch the frog's mouth.
2. Label the following structures:

A. *Upper Jaw*
Maxillary Teeth   Internal Nares
Vomerine Teeth   Eustachian Tube Openings

B. *Lower Jaw*
Tongue   Glottis
Pharynx   Esophagus

Assign these questions:

1. What is unusual about the tongue attachment? It is attached in front.
2. Where does the glottis opening lead? To the lungs.
3. Where do the Eustachian tube openings lead? To the ear cavity.

4. How do the maxillary teeth aid the frog? Helps the frog hold, crush, and grind food.
5. What is the purpose of the vomerine teeth? To hold food.
6. Where do the internal nares lead? To the external nares (respiration).
7. How does a large mouth benefit the frog? Makes it easier to catch prey.
8. Do female frogs have a vocal sac? No. Only males.

ACTIVITY 157

Have pupils remove the brain and lay it in a tray. Offer these suggestions:

1. Locate, inspect, sketch and label the following structures:

Olfactory Lobe	Cerebellum
Cerebrum	Medulla
Optic Lobe	Spinal Cord

ACTIVITY 158

Tell pupils to lay the frog on its dorsal side, pin down each limb, and make an incision from the neck to the anus. Carefully remove the skin and muscles which cover the internal organs.
Remove each of the following structures.

Stomach	Gall bladder
Intestine	Pancreas
Liver	Lungs

Have pupils examine stomach contents and inspect intestinal walls.

ACTIVITY 159

Have students remove the heart and lay it in a tray. Tell them to locate, inspect, and sketch these structures:

Front View	*Back View*
Right Auricle	Pulmonary Veins
Left Auricle	Sinus Venosus
Ventricle	Left Auricle
Conus Arteriosus	Right Auricle
	Vena Cava

List the following questions:

1. How does the frog heart differ from the fish heart? The frog has a three-chambered heart; the fish has a two-chambered heart.
2. What structure surrounds the heart muscle? Pericardium.
3. What is the purpose of arteries? Carry blood away from the heart.
4. What is the function of the veins? Carry blood to the heart.
5. How does blood enter and leave the heart? Blood travels from the sinus venosus into the right auricle. Blood from the lungs enters the left auricle. Both auricles contract and send blood into the ventricle. Blood leaves the ventricle and moves to different parts of the body.

FOURTEEN SUGGESTIONS FOR FURTHER STUDY

Assign these topics to students who wish to do extra work:

1. Try the following experiments on a live frog:
   a. Move an object toward the frog's eye. How does the frog react?
   b. Gently pull the digits on the hind limb with forceps. If the frog doesn't respond, try pulling harder. How does the frog react?
   c. Watch the frog's throat. How many times does it move each minute?
   d. Encourage the frog to jump. After several jumps count the frog's breathing rate. How does it compare with your observation of Part C?
2. Remove the heart of a freshly killed frog. Transfer to a beaker which contains saline solution. How many heart beats do you count per minute?
3. Repeat Topic #2. Work with a partner. Place one heart in a warm saline solution. Place another heart in cold fresh water. Count the heart beats per minute for 10 minutes. Compare results.
4. Dissect the hind limb of a frog. Tease away the skin and muscle tissue. Locate the sciatic nerve. What is the function of the sciatic nerve?
5. How do frogs respond to vibration? Place a frog in a plastic or glass container. Set a vibrating fork at different areas around the container. Note the response.
6. How do frogs respond to tactile stimulation? Touch different parts of the frog's body with various stimuli, e.g., wooden splint, nail, glass rod, finger, etc. Note the response. Does the frog respond more to a specific stimulus? Is one part of the body more sensitive than another?

7. Explain how a frog breathes. Trace a breath of air through the respiratory system. List the structures involved in the breathing process.
8. The frog has 10 pairs of cranial nerves. List them and give their function.
9. How does a frog's reproductive system compare with a bony fish? How is it alike? How does it differ?
10. Examine a frog skeleton. Find the following bones:

Head	*Body*	*Front Limb*
Maxilla	Clavicle	Phalanges
Occipital Condyle	Supra Scapula	Metacarpus
Squamosal	Vertebrae	Carpus
Fronto-Parietal	Atlas	Radio-Ulna
Hyoid	Coracoid	Humerus
Nasal	Urostyle	*Hind Limb*
Vomer	Ilium	Femur
Pterygoid	Ischium	Tibio-Fibula
	Pubis	Tarsus
	Sternum	Metatarsus
		Phalanges

11. Examine a fish vertebrae. Locate these structures: Pleural ribs, centrum, haemal arch, neural arch and neural spine.
12. Trace a piece of food through a grasshopper's body. Tell what happens as the food passes through each digestive structure. Do the same for an earthworm, clam, and fish.
13. Compare the nervous system of an earthworm, clam, crayfish, fish, and frog. List the similarities and differences.
14. Report on methods used to kill the following harmful insects: Cockroach, grasshopper, aphids, scale, cotton boll weevil, tomato worm, and clothes moth.

chapter ten

16 ACTIVITIES THAT WILL HELP STUDENTS LEARN ABOUT ECOLOGY

OVERVIEW

Students examine plant-animal communities and how they interact. The role of producer, consumer, and decomposer is discussed. Pupils take random sample counts and construct a food chain sequence. Student games—*Meal Wheel, Lion's Share,* and *Survival*—stress the food chain concept, competition within a community, and land destruction. The last section emphasizes pollution. The final activity allows pupils to coordinate their artistic talents and produce limericks, poems, or cartoons depicting ecological events.

Begin discussion by drawing 6 different shapes on the chalkboard, e.g., circle, triangle, rectangle, square, cross, and disc (Figure 55). Tell pupils each shape represents a different community of living organisms.

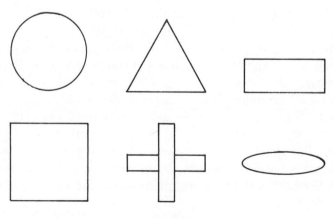

Figure 55

1. What makes up a certain population? Answers will vary.
2. What determines the shape or distribution of a community? Answers will vary.
3. What factors might alter the size and shape of a community? Answers will vary.

Emphasize communities involve all the plant and animal populations within a certain area. Individuals interact to form populations. Populations join hands to form communities. And communities, along with the environment, food supply, and weather conditions make up the ecosystem.

Tell students to bring a deck of regular playing cards to class.

HOW TO MEASURE AND RECORD DIVERSE POPULATIONS

How many organisms live in a particular area? If only a few exist, a person can easily count each member of the population. However, it would be impractical to use this method for counting large populations.

Random sampling, also called chance sampling, eliminates bias in selection. Every member of a population stands an equal chance of being counted.

ACTIVITY 160

Give each student an 8 by 8 inch piece of plain white paper. Relate this story: A population of holes punched into town. They spread everywhere, too many to count. How can we estimate the number of holes? By taking a population sample count.

Offer these guidelines:

1. With a ruler and pencil, divide the paper into 64 square inch quadrates. Number each quadrate (Figure 56).
2. Turn the paper over. Punch holes throughout the paper with a pencil. Count the number of holes as you punch them. Record this number.
3. Number 1 through 25 on a separate piece of paper. Randomly select 25 numbers from the quadrate paper. Place these selections, one at a time, beside each listed number (See Figure 56).
4. Turn the quadrate paper over. Count the holes within each selected quadrate. Include those touching the line. Write this number beside the selected quadrate on the numbered list (See Figure 56).

5. Add up the total number of holes counted. Divide the total number by 25. Multiply by 64. The answer will be the random population count. How does the sample count compare with the actual count?

   Example:

a. Total for 25 random quadrates = 54
b. Divide 25 into 54 = 2.16 average per quadrate
c. 2.16 x 138 total random count
d. Actual total number of holes = 153
e. How does the sample count compare with the actual count? The random count fell 15 short of the actual count.

Figure 56

Suggested Problems And Questions

1. Repeat Activity 160. Instead of picking random quadrates, have pupils bend a paper clip to form a square inch (Figure 57). Tell each student to toss the clip on the quadrate paper 25 times.
2. Repeat Activity 160. Instead of picking random quadrates, give students small rubber bands approximately one inch in diameter. Tell each pupil to toss the rubber band on the quadrate paper 25 times.
3. Have students tape square inch quadrate sections on their laboratory tables. Then scatter paper punch or rice over the quadrates. They can toss string loops, paper clips, or rubber

bands over the table. To insure complete unbiasness, have pupils face away from the table and make the toss over their shoulders.
4. What factors might make a person bias in counting? He might pick samples which are more physically attractive, e.g., taller, rounder, longer, etc.
5. Why should random sampling be used in large populations? It is the only practical way to get an estimated proportion of the total population.
6. Pass out filter discs or have students make their own. Tell them to mark off equal sections on the discs and number each section. Then pepper the disc with dots. Place a pin in the center of the disc (Figure 58). Spin the disc several times. Count the section which stops nearest you. Average the number of dots counted for each disc. How does the sample count compare with the actual count?

Figure 57

OBSERVING HOW COMMUNITY MEMBERS RELATE

Ask pupils for their definition of ecology. List their responses on the blackboard. Underline the key items. Write a new definition using the key words.

Set up Activity 168, hay infusion.

Discuss the relationship between community members. Emphasize the following points:

1. Producers (plants) convert solar energy to chemical energy through the process of photosynthesis.

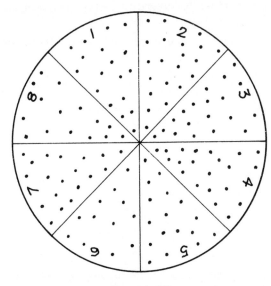

Figure 58

2. Consumers—The animals which feed upon plant eaters or the plants themselves.
3. Decomposers—Bacteria, fungi, and small animals which break down waste material and dead organisms.
4. The interaction between producers, consumers, and decomposers make up the ecosystem. This includes energy transfer throughout the entire system.

ACTIVITY 161

List these items on the blackboard: Bacteria, mud, water, water lily, water bug, toad, water snake, minnow, heron, alligator, and man.
Have pupils:

1. Show, with arrows and diagrams, how these living organisms relate to one another, i.e., how does energy transfer from one organism to another?
2. Tell what can happen to destroy the energy flow.
3. Name the producers, consumers, and decomposers.

Assign pupils these additional questions:

1. Honey bees interact with flowers. Bees make honey from pollen and help pollinate flowers. What would happen if the flowers suddenly disappeared? The bees would seek a new community.
2. Living organisms die and replenish the soil. What would happen

if organic soil became contaminated? The community would eventually die off.

3. List some animals which have limited distribution. Penguins, polar bears, coral, caribou, desert rats, and so on.

4. Why did the dinosaurs die off? No one knows for sure. Some theorists say the herbivores ate all the plants, died off, and deprived the meat-eaters of food. Others claim various diseases, small egg-sucking mammals, competition, or changing geologic structure destroyed them.

5. Why are some plants and animals restricted to certain habitats? Climatic factors and physical structure of organisms help determine where the organism will live.

WAYS TO STUDY FOOD CHAINS

Discuss the complexity of food chains or webs which exist in every community. Stress how leaks occur in the food chain. For example, write this sequence on the blackboard:

Protozoa   Diving Beetle   Nymphs (insect larvae)
Bass   Man

Relate not all diving beetles are eaten by nymphs or all nymphs eaten by bass or every bass caught and eaten by man. Some of these animals die of old age or become eaten by other species.

ACTIVITY 162

This activity reinforces the concept of food chain complexity. Give each student scissors, glue, colored and plain white paper. Have them use the information chart (Figure 59) to construct a food chain sequence. Tell them to cut out a colored square inch piece of paper to represent each organism. Then paste each square on white paper (Figure 60). Draw arrows to show how the organisms interact.

Offer students these additional problems:

1. What would happen if:
   a. Green disappears from the community?
   b. Blue disappears from the community?
   c. White disappears from the community?
2. Construct 3 different plans. Rearrange the symbols each time. Do these changes affect the community?

Mention how living organisms stay within their habitat, or niche. They relate to environmental conditions, food supply, and other members of the community—both friends and enemies.

Food Chain Chart

ORGANISMS	FOOD CHAIN SEQUENCE
WHITE	EATS GREEN. EATEN BY YELLOW, ORANGE AND BROWN.
BLUE	EATS GREEN. EATEN BY ORANGE, YELLOW AND BROWN.
YELLOW	EATS BLUE AND GREEN. EATEN BY BROWN AND YELLOW.
ORANGE	EATS WHITE, BLUE AND YELLOW. EATEN BY BROWN.
BROWN	EATS YELLOW, WHITE, BLUE AND ORANGE. NOT EATEN.
GREEN	EATS NOTHING. EATEN BY YELLOW, BLUE AND WHITE.

Sample Plan:

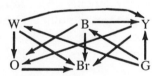

Figure 59

Eats ⟶ Eaten

Figure 60

ACTIVITY 163

Divide students into small groups. Tell each group member to bring a tree branch (or small log) covered with plant material and small organisms to class.

The laboratory session includes macroscopic and microscopic examination. Provide reference books, dissecting tools, and hand lens or microscopes. Have pupils make a thorough investigation before transferring pertinent information to an observation chart (Figure 61).

A discussion preluding the laboratory investigation starts the ball rolling. Here is an excellent "attention getter"—write the following verse on the blackboard and ask for a volunteer to supply the missing words.

> A tadpole swimming through the grass
> was eaten by a baby bass.
> The bass was happy with his meal,
> But soon was swallowed by a teal.
> And as the teal flew toward the sun,
> He was felled by a_____.

Any sharp, ecology minded student with imagination (and helpful hints from the teacher) quickly sees "hunter's gun" nicely completes the last sentence. The poem points up two things: survival of the fittest and man's dominance over lesser beings.

Give students the following questions. Review them in post laboratory discussion.

1. Is each species independent, or do some species share the same position in their environment? Answers will vary.
2. Can two or more species indefinitely continue to occupy the same ecological niche? Yes, so long as the food supply lasts.
3. How can a dominant species affect conditions of environment for other populations? A dominant species may set up territories, drive off competitors, and control the food supply.
4. What factors limit the number of niches which may be occupied? Environmental conditions, climate, competition, and available food supply.

ACTIVITY 164

Take the class on a walk around campus. Visit three (optional) different areas, e.g., football field, hedges along the cafeteria, front lawn, etc. Have them describe the physical environment and list the plants and animals which live in each niche.

Observation Chart

	What is it?	How many?	Where located?	What doing?	Size (MM)	Shape	Sketch
PLANTS							
ANIMALS							

Figure 61

Have pupils answer these questions:

1. How are the habitats different? How are they alike? Answers will vary.
2. List the consumers, producers, and decomposers for each habitat. (Students must assume bacteria and other small organisms live in the soil.) Answers will vary.
3. Design an ecosystem for all three niches. (Remind students an ecosystem consists of all living organisms, environment, food supply, and weather conditions.) Answers will vary.

Go over responses in class.

Ask students how anatomical structure and specific needs determine where an organism lives. For example, the Koala bear of Australia feasts on eucalyptus leaves. Hence, countries without eucalyptus trees are devoid of Koala bears. A hawk's beak would be unsuitable for probing in the mud; a duck would have a rough time clinging to a branch, and so on.

ACTIVITY 165

Give students a niche chart (Figure 62). Provide reference books or materials to help them complete their charts. The chart shows a particular structure. The student identifies the organism's structure, habitat, and how the structure is suited for the environment.

ACTIVITY 166

Meal Wheel: A Student Game Emphasizing Food Chains

There's nothing like a game to keep interest flowing. And *Meal Wheel* reinforces the food chain principle—Living things are directly linked to one another by what they eat.

Give pupils these guidelines:

1. Cut out a circle (5″ diameter) from an index card.
2. Mark off the card in 8 equal sections.
3. Write these words on the cardboard wheel, one word per section: plant, grasshopper, frog, snake, hawk, fox, man, and pollution.
4. Push a pencil through the center of the wheel. Rotate the pencil to widen the hole. The wheel should spin freely around the pencil (Figure 63).
5. Objective: To turn the wheel until all organisms in the food chain appear.

Here are the rules for *Meal Wheel:*

1. Choose a partner. Each person takes a turn spinning the wheel.
2. The player spinning holds a pencil in his hand with his thumbnail facing him. The thumbnail represents a pointer. He writes the name of the section on a piece of paper which comes to rest over his thumbnail. If the wheel stops between sections, the subject spins again.
3. The spinning player writes down the name of each section only once.
4. Each time pollution appears, the spinning player loses one of his previous selections. For example, the player has written down

Niche Chart

STRUCTURE	ORGANISM	HABITAT	HOW IS STRUC-TURE SUITED FOR ENVIRON-MENT
	(Shore bird beak)	(Shoreline)	(Long, straight bill for probing into sand)
	(Feet of water and shore birds)	(Water)	(Primarily for swimming or paddling)
	(Hind leg of tree frog)	(Leaves, trees, water plants)	(Skin color blending. Discs on toes aid climbing)
	(Tentacle of squids, octopuses and nautiluses	(Marine environment)	(Grasping prey and movement)
	(Rear leg of grasshopper)	(Open fields and gardens)	(Jumping)
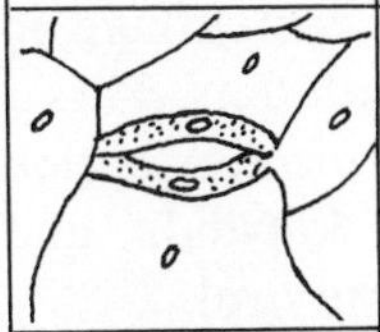	(Stoma of a leaf)	(Plant community)	(Opening on the epidermis of a leaf or stem. Gas exchange)
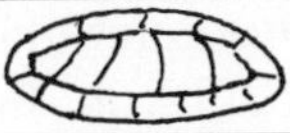	(Turtle shell)	(Water)	(Protection)
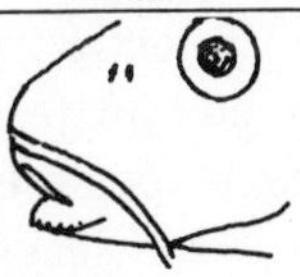	(Fish mouth - carp)	(Fresh water)	(Sucking food from mud bottom)

Figure 62

grasshopper, fox, and snake. He spins pollution. He must remove one of these selections.

5. Repeat several times. The student who completes his list first wins the game.

Figure 63

Demonstrations And Experiments

1. Demonstrate examples of animal mimicry and camouflage in nature. Hold up a piece of white paper covered with dots. Pass a smaller piece of dotted paper in front of the larger piece. Mention how some animals, e.g., flounders, chameleons, katydids, and walking sticks blend in with their environment for survival.
2. Hold up pictures of a monarch and viceroy butterfly. Relate how the viceroy butterfly mimics the monarch. Some birds find monarch butterflies distasteful. The viceroy seeks protection by looking like the monarch. Pass around preserved butterfly specimens if you have them. Let students examine each specimen.
3. Does aquarium size affect fish growth? Find out. Place several equal sized minnows in two different aquariums. One aquarium should be twice as large as the other. Keep conditions the same for all fish, e.g., water temperature, water level, food supply, oxygen level, etc. Make periodic checks of fish growth.
4. How do crowded conditions affect growth? Plant 20 or more beans in one pot; plant 3 or 4 beans in an identical pot. Fill both pots with the same soil. Give each pot the same amount of water, light, and temperature. Check periodically. What are your conclusions?

5. Locate a plot of ground near the school cafeteria. Sketch the area. Include paper debris, orange peelings, metal tin tabs, sticks, rocks, and so on. Visit the area once a week for nine weeks. Make a new sketch each visit. Record the changes which take place each week.
6. Find an empty lot. Scatter old rags and boards around. Will this debris provide a new home for organisms? Make periodic checks to find out.
7. Fill three trays with the following ingredients:

Tray A	*Tray B*	*Tray C*
Broken sticks	Sandy or clay soil	Compost, lawn clippings, or leaves
Small rock fragments	Paper strips	

Set the trays in an undisturbed area around home or school. Check the trays every 2 or 3 days. Have they attracted any organisms? Which tray, if any, attracts the most organisms? How do you account for this? Would adding food to each tray entice more organisms? Try it.

LION'S SHARE: A STUDENT GAME STRESSING OVER-POPULATION

Mention how different populations compete in many ways for survival. They compete for living space, food, water, light, and so on. An environment, left undisturbed, may succeed for years. One exposed to changing natural forces or pollution can reduce or destoy a living population.

ACTIVITY 167

Let students play *Lion's Share,* a card competition game. Situation: Four living organisms—A, B, C and D—arrive on a deserted island. There's very little food available. Who will get the lion's share? Who will be left with only scraps?

Here is the procedure:

1. Students work in pairs. Each group needs a regular deck of playing cards.
2. Each player chooses two "island organisms," e.g., A and C, D and B, B and A, etc.

3. Write the *Lion's Share* point chart on the blackboard. Include a competition bracket chart (Figure 64).
4. Have players copy the bracket chart on a piece of paper. Each player lists both selected letters on a bracket opposite his opponent's choice. For example, Player #1 picks A and C; Player #2 likes B and D. The competition bracket looks like this:

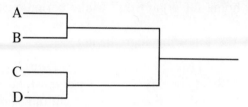

5. Shuffle the cards. Place them face down on the table. Each player, in turn, removes the top card and compares the card against the chart to find its point value.

Lion's Share Point Chart

	HEARTS	CLUBS	DIAMONDS	SPADES
A	0	3	-1	2
K	1	-1	-2	0
Q	2	4	0	6
J	-3	0	3	5
10	4	-2	-1	3
9	3	5	2	0
8	-4	0	1	-4
7	0	-3	-3	0
6	4	5	4	-5
5	1	-2	-1	0
4	-1	1	2	1
3	5	4	2	3
2	6	3	0	2

Figure 64

Competition Bracket Chart

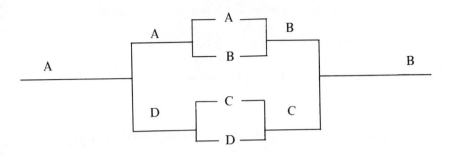

Figure 64 B

6. Each player turns over 5 cards and adds the points algebraically to determine the point total. If a tie occurs, each player continues to draw until one wins. The winning letter moves to the next bracket.
7. Bracket play continues until the *Lion's Share* or champion is determined. The first round losers play off for third position and consolation points.
8. Award points in this manner:

> *Lion's Share* = 10 points
> Second Position = 7 points
> Third Position = 4 points
> Consolation = 3 points

The player who receives the most points dominates the island:

Example:

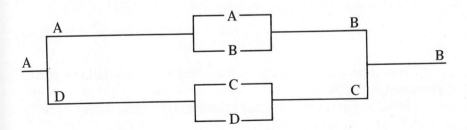

Points Awarded:

Lion's Share — B (10 points)
Second Position — C (7 points)
Third Position — A (4 points)
Consolation — D (3 points)
Player #1 — 11 Total Points
Player #2 — 12 Total Points

Player #2 controls the island.

EXAMINING THE EFFECTS OF OVERPOPULATION

ACTIVITY 168

Set up a hay infusion. Mix pieces of hay with water. Boil for approximately 10 minutes. Set aside for 3 days. Add 200 ml of creek or pond water. Leave stand for 5 or 6 days.

Have pupils make daily microscopic observations. Provide the following guidelines:

1. Make 3 microscopic slides. Label the slides A, B, and C. Use a different dropper for each slide. Each letter represents a different niche.
2. Sketch the organisms present in each niche. Which organisms seem to dominate? List the various ways these organisms move around.
3. Make daily observations for the next 5 days. What changes occur? Do new organisms appear? Answers will vary.
4. What organisms seem to dominate each slide? Are these organisms eating the Lion's Share of food? How can you tell? Answers will vary.
5. How long will the organisms remain alive? How may poisons or pollutants affect the entire community? Where might these toxins come from? Answers will vary.
6. How do these organisms compete for food, and shelter? List specific examples. Answers will vary.

ACTIVITY 169

How does the availability of space influence population size and competition for food?

Have pupils transfer the hay infusion into 5 separate test tubes in the following manner:

Test Tube Number	Hay Infusion (ml.)
1	10
2	20
3	30
4	40
5	50

Let pupils make a microscopic slide from each test tube. Have them answer these questions:

1. Does the organism population change or remain the same each day? Answers will vary.
2. Does the 50 ml infusion contain 5 times as many organisms as the 10 ml. infusion? Answers will vary. Does the 30 ml infusion contain 3 times as many organisms as the 10 ml infusion? Answers will vary.
3. Which test tube infusion appears to hold more organisms? Which test tube infusion appears to offer more food and protection? Answers will vary.
4. Set the test tubes in a safe place for 2 or 3 weeks. Reexamine each infusion. Have any changes taken place? Summarize your findings. Answers will vary.

SURVIVAL: A STUDENT GAME ON POLLUTION

Ask students what balance in nature means. Bring up the idea a plant-animal community will remain the same, year after year, if left undisturbed. In some cases, plowing the land and removing vegetation will drive animals from their homes. This creates an upset or imbalance in nature. Floods and storms carry away valuable top soil; parasites spread diseases which kill both animals and plants. Man and nature play a dual role—both preserve; both destroy.

ACTIVITY 170

Have pupils play *Survival,* a 16-card game. Here is the procedure:

1. Pupils work in pairs. Remove these cards from a regular deck: 4 Aces, 2 Jokers, 4 Dueces, King of Diamonds, Queen of Spades, Queen of Hearts, 9 of Spades, 8 of Clubs, and 3 of Hearts.
2. Make a community chart approximately 8 square inches (Figure 65).

Community Chart

SOIL	INSECTS
BACTERIA	BIRDS
PLANTS	MAMMALS

Figure 65

3. Players choose sides by flipping a coin. One selects *black* and becomes *Acre Taker;* the other chooses *red* and becomes *Acre Saver.*
4. Each player uses a different colored marker—black or red (optional)—to put on the community chart. Small plastic checkers or beans make adequate markers.
5. Shuffle the cards. Player One, *black,* begins play by removing the cards, one at a time. He checks their value against the chart (Figure 66). Player Two, *red,* places the markers on the chart.

The rules are as follows:

1. *Black* removes a card and lays it face up on the table. *Red* moves the marker accordingly.
2. Aces and deuces belong to *black.* An ace or deuce means a black marker can go into any empty space on the community chart. If all spaces are filled, the marker can replace a red marker anyplace on the board.
3. All cards listed on the value chart, including the Joker, belong to Red. Red markers must fall on the drawn-listed card only. For example, if *red* receives the Queen of Spades, he must place a red marker on the bird section. If he receives the 8 of Clubs, he must place a red marker on the bacteria section, and so on (See Figure 66).
4. If the chart is full and *red* receives the joker, he can remove a black marker, replacing it with a red marker.
5. If *black* has a marker on a listed section, say mammals, and red draws the same section, he replaces the black marker with a red marker.

Value Chart

PLAYER TWO	SOIL	3 OF HEARTS
	BACTERIA	8 OF CLUBS
	PLANTS	9 OF SPADES
	INSECTS	KING OF DIAMONDS
	BIRDS	QUEEN OF SPADES
	MAMMALS	QUEEN OF HEARTS
PLAYER ONE	FREE MOVE	JOKER
	BLACK MOVE	ACE
	BLACK MOVE	DEUCE

Figure 66

6. *Red* has a slight disadvantage. If his marker covers a section, say insects, and he redraws the King of Diamonds, he gains nothing. Also, *black* can move anywhere on the chart. *Red,* with the exception of the Joker, must follow the drawn-listed card.
7. Participants continue to play until one player fills the chart with his markers. If cards run out, reshuffle and continue the game. If *black* wins, the community dies. If *red* wins, the community survives.
8. Players trade off between moving markers and drawing cards.
9. There is no time limit. Let students play until they lose interest. Have them keep their score sheets. Tabulate all the score sheets. Who wins the most—black or red? Why would black hold a slight edge? How does this game relate to living conditions?
10. If some pupils are without partners, let them play alone. They can easily draw cards and move the markers.

WAYS TO TEST FOR POLLUTION

Students are quite familiar with the plight of conservationalists and ecologists trying to save the land, air, and waterways of the world.

Arouse pupil interest through films, filmstrips, slides, photographs, or guest speakers. Emphasize these conditions: Synthetic suds crowding a drainage canal, effects of radioactivity, industrial wastes, and smoke from incinerators. Stress how pollution destroys plant and animal communities.

Discuss the effects of air pollution. Mention how air pollution damages property and health. Allow students to make brief reports to the class. Environmental issues appear daily in the paper.

ACTIVITY 171

Have pupils collect samples of water from any of these areas: running streams, stagnant ponds, muddy pools, faucets, rivers, lakes, street gutters, irrigation ditches, etc.

Give them these guidelines:

1. Test each water sample. Evaporate 20 ml of each solution in a dry, clean beaker.
2. Check the inside walls and bottom of each beaker. Examine scrapings under a hand lens or microscope.
3. Answer the following questions:
   a. Which beaker left the greatest amount of residue? Answers will vary.
   b. How are the residues alike? How are they different? Answers will vary.
   c. Are any of these residues harmful? How can you be sure? Answers will vary.
   d. Do you think boiling may have killed any harmful products? Answers will vary.
4. Sketch the residue from each beaker. List the physical properties, e.g., color, shape, texture, etc.

ACTIVITY 172

Let pupils trap dirt particles, chalk dust, etc., in the classroom. Give these directions:

1. Place scotch tape lengthwise along a microscopic slide, sticky side up.
2. Wrap the ends of the slide with strips of masking tape. This will hold the scotch tape in place.
3. Prepare 2 or 3 slides. Label them. Examine each slide under a microscope. Describe the foreign particles which cling to each slide.
4. Place each slide in a different location throughout the classroom.

5. Make daily observations. Do new substances appear? Can you identify them?
6. What will happen if the slides are left alone for 3 or 4 weeks? Try it.

ACTIVITY 173

Have students trap outdoor floating particles. Offer these directions:

1. Fill the bottom of 2 or 3 Petri dishes with glycerin.
2. Place containers in different locations, e.g., near cafeteria, near bus garage, near an incinerator, football field, etc. Set containers where air pollution seems the greatest.
3. Check containers every 4 or 5 days. Particles may be difficult to identify. Refer to *A Key To The Identification Of Air Pollution Particles*. Science Activities Magazine, October, 1973, pp. 29-31.

ACTIVITY 174

Chemicals plague some water systems producing hard water. Hard water leaves a scum in pipes, sinks, bath tubs, and shower walls. Let students test for the presence of hard water.

Part A.

Add 10 grams of Epsom salts to 100 ml's of water. Mix thoroughly. Evaporate 20 ml's of solution. Does any residue remain? Heating will not drive away the Epsom salts.

Part B.

Add 10 grams of calcium bicarbonate to 100 ml's of water. Mix thoroughly. Evaporate 20 ml's of solution. Does any residue remain? No. Heat drives away the calcium.

Part C.

Add 10 grams of crushed gypsum (or ground chalk) to 100 ml's of water. Mix thoroughly. Evaporate 20 ml's of solution. Does any residue remain? Heating will not drive away the gypsum.

CREATING ECOLOGICAL GRAFFITI

ACTIVITY 175

Let the students bring the unit to a close by writing limericks, short poems, or creating cartoon sketches on ecology themes. Display the final products on the bulletin board.

Examples:

A radiologist named Jack Clark
swallowed an isotope unmarked.
He became infuriated
for becoming irradiated,
but smiled as he glowed in the dark.

A stream flowed once unpolluted.
It now has its bottom extruded.
The fish complain not,
they're still being sought
in an environment surely well-sooted.

TWELVE SUGGESTIONS FOR FURTHER STUDY

Assign these items to interested students:

1. What are "Living Fossils"? What do they infer regarding survival?
2. Refer to a biology textbook. Try the suggested photosynthesis experiments.
3. Find examples of these animal behaviors: Mating rituals, aggression, feeding, and establishing territorial rights. How do these behaviors lead to survival?
4. List several omnivores, carnivores, and herbivores. How do they interact with each other?
5. How does the carbon cycle and nitrogen cycle help the flow of energy reach every organism?
6. What is a parasitic chain? How does it differ from a food chain?
7. Name some plants and animals which are becoming extinct. Tell what conditions lead to the extinction of some organisms.

8. Throw 30 beans across a lawn. How many will germinate? What factors inhibit growth? What factors help growth?
9. Bury 10 beans in a pile several inches beneath some top soil. How many sprout? Bury 10 more beans several inches beneath some top soil. Keep them separated. How many sprout? What do you conclude about this experiment?
10. Make several food chains. Use arrows to show the flow of energy.
11. Name several ways man has altered the balance in nature. How has he attempted to control imbalance?
12. Place scotch tape slides and glycerin containers in different areas around town. Check periodically. Try to identify the captured particles (See Activity 173).

REFERENCES

Bishop, Lewis, Bronaugh, *Focus On Earth Science,* Charles E. Merrill Publishing Co., Columbus, Ohio, 1969, p. 325.

Brice, James C., *Laboratory Studies In Geology* (201), W.H. Freeman and Company, San Francisco, 1962, pp. 7, 8.

Eardley, A.J., *General College Geology,* Harper & Row, New York, 1965, pp. 465, 468.

Earth Science Curriculum Project, *Investigating The Earth,* Houghton Mifflin Company, Boston, Massachusetts, 1967, pp. 363-364.

English, George Letchworth, *Getting Acquainted With Minerals,* Chapter 5: McGraw-Hill Book Co., Inc., New York, 1934, P. 32.

Fenton and Fenton, *The Fossil Book,* Doubleday & Company, Inc., New York, 1958, pp. 1-3.

Geology And Earth Science Sourcebook, Holt, Rinehart and Winston, Inc., New York, 1962, pp. 11, 27-28, 71.

Joseph, Brandwein, Morholt, Pollack, Castka, *A Sourcebook For The Physical Sciences,* Harcourt, Brace & World, Inc., New York, 1961, pp. 33, 44, 146-147, 250, 308.

Ramsey and Burkley, *Modern Earth Science,* Holt, Rinehart, and Winston, Inc., 1965, pp. 169, 214, 433, 435.

Schneider, Herman and Nina, *Science And Your Future,* D.C. Heath and Company, Boston, 1961, p. 209.

Science Service, *Earthquakes,* Nelson Doubleday, Inc., p. 35.

Smith/Lisonbee, *Your Biology,* Harcourt, Brace and World, Inc., New York, 1962, p. 24.

Thurber, Walter A., and Kilburn, Robert E., *Exploring Life Science,* Allyn and Bacon, Inc., 1966, pp. 126, 128.

Thurber, Walter A., and Kilburn, Robert E., *Exploring Earth Science,* Allyn and Bacon, Inc., 1965, pp. 107, 218, 220-222, 232, 255, 278, 422.

Hoehn, Robert G., *Science Activities,* The Log Squad, Volume 6, No. 5, January, 1972, pp. 35-36.

.... *Science Activities,* Map-Making As "Fun Stuff," Volume 8, No. 1, September, 1972, pp. 10-12.

.... *Science Activities,* A Griddle Earth, Volume 8, No. 4, December 1972, pp. 14-16.

••••••••••••••••• *Science Activities,* Fossilizing, Volume 10, No. 1, September, 1973, pp. 22-24, 44, 45.

••••••••••••••••• *Science Activities,* The Paper Rocket, Volume 8, No. 5, January, 1973, pp. 16-18.

••••••••••••••••• *Science Activities,* Planet Enigma, Volume 9, No. 1, February, 1973, pp. 16-18.

••••••••••••••••• *Science Activities,* Volcanoes: Part II - Experimental Studies, Volume 9, No. 2, March 1973, pp. 19-21, 47.

••••••••••••••••• *Science Activities,* Space Chips, Volume 9, No. 3, April, 1973, pp. 10-12.

••••••••••••••••• *Science Activities,* Tektites, Anyone?, Volume 9, No. 4, May, 1973, pp. 16-17.

••••••••••••••••• *Science Activities,* "Punfun," Volume 10, No. 2, October, 1973, p. 41.

Sherman, Ruth G., *Science Activities,* The Chronicles of Chromobacter, Volume 8, No. 4, December, 1972, pp. 10-13.

Index